T0331643

THE ENERGY
CONUNDRUM

Climate Change, Global Prosperity, and the
Tough Decisions We Have to Make

THE ENERGY CONUNDRUM

Climate Change, Global Prosperity, and the
Tough Decisions We Have to Make

Neil A.C. Hirst

The Grantham Institute, Imperial College London

World Scientific

NEW JERSEY · LONDON · SINGAPORE · BEIJING · SHANGHAI · HONG KONG · TAIPEI · CHENNAI · TOKYO

Published by

World Scientific Publishing Europe Ltd.

57 Shelton Street, Covent Garden, London WC2H 9HE

Head office: 5 Toh Tuck Link, Singapore 596224

USA office: 27 Warren Street, Suite 401-402, Hackensack, NJ 07601

Library of Congress Cataloging-in-Publication Data

Names: Hirst, Neil, author.

Title: The energy conundrum : climate change, global prosperity, and the tough decisions
  we have to make / by Neil Hirst (Imperial College London, UK).

Description: New Jersey : World Scientific, [2018]

Identifiers: LCCN 2017043811 | ISBN 9781786344601 (hc : alk. paper)

Subjects: LCSH: Energy policy.

Classification: LCC HD9502.A2 H54 2018 | DDC 333.79--dc23

LC record available at https://lccn.loc.gov/2017043811

British Library Cataloguing-in-Publication Data

A catalogue record for this book is available from the British Library.

For any available supplementary material, please visit
http://www.worldscientific.com/worldscibooks/10.1142/Q0135#t=suppl

Desk Editors: Dr. Sree Meenakshi Sajani/Jennifer Brough/Koe Shi Ying

Typeset by Stallion Press
Email: enquiries@stallionpress.com

Printed in Singapore

Foreword

I am delighted to recommend Neil Hirst's insightful book on energy policy, *The Energy Conundrum*. In a complex and rapidly changing energy scene, this comprehensive account of global energy policy is particularly welcome. At its heart is a critical question facing humanity, "how can we meet our growing need for affordable and secure energy services while also limiting climate change to acceptable levels?" The emphasis is on fundamental understanding of the full range of moral, political, social, technical, and economic issues as they appear in different parts of the world, poor as well as rich.

In recent decades, increased energy supply has sustained unprecedented improvements in living standards in many parts of the world. We should celebrate that. Now, policy makers have to find ways to continue and broaden the process while curtailing carbon emissions and local pollution. They also need to bring modern energy supplies and safe cooking facilities to the billions of people who are still without them.

Even in the industrialised economies, energy has to be affordable as well as clean. Fortunately, this is increasingly possible. Improvements in energy efficiency are reducing costs as well as emissions. Environment-friendly renewables are increasingly competitive with traditional fossil fuels and we are finding smarter ways to integrate variable renewable supplies into our energy systems.

But our problems are far from being solved. The impressive progress being made in the decarbonisation of electricity is not matched by similar progress with transport or heat. We need to step up research and development on advanced energy sources and systems. Specific support for the deployment of low carbon generation is still important. But we need to move progressively towards more market-based incentives that bring into play the full range of emissions-saving options and give clear signals to business investors.

Meeting the world's energy needs will generate immense opportunities for trade, investment, economic development, and jobs in the coming decades. The IEA's growing family is committed to policies that keep the avenues for these flows open.

Energy security is still vital. We will continue to depend on fossil fuels at least for some decades to come. The political stability of some energy-producing countries in the Middle East and elsewhere continues to be a concern. There is a rapid increase in shipments of natural gas in liquefied form and this trade now raises similar security questions to those of oil. Cyber security is also a growing worry.

We are definitely not on course to meet the environmental and social objectives of many countries' energy policies. But we may be approaching a turning point. The 2015 Paris climate summit has given a powerful lead. We must build on that. The progress of clean energy, especially solar and wind power, is spectacular and may soon be complemented by advances in energy storage. China, which is the largest energy consumer, is coming to the end of an era of rapid industrialisation and is turning to a cleaner and less energy-intensive pathway. Global trends on energy efficiency have improved. IEA analysis shows that over the last 2 years, for the first time except during periods of recession, the growth of world carbon emissions has come to a halt, although the global economy grew.

The book rightly emphasizes that international cooperation, especially between industrialised and emerging economies, will be essential for setting the world on a sustainable energy path. The IEA has recently taken steps to open its doors to the emerging economies, with several — including China, India and Indonesia — now Association members of the Agency. As these countries become ever more important in global energy markets, we need them on board as distinguished members of the IEA family as we tackle the world's energy challenges. In addition, increased dialogue between governments and industry is also critical for finding solutions. Thus, the IEA has actively engaged with companies to share views on such key issues as investment and the impact of digitalisation on the energy sector.

I am optimistic that we can solve this energy conundrum — to use Neil Hirst's terminology — but it remains a tough challenge. With industrialised and emerging economies working together around the table, I am convinced that the IEA can make an important contribution. For anyone working in the energy field, whether in government, business, academia, or NGOs, who needs to understand the depth and complexity of this challenge, *The Energy Conundrum* is essential reading.

Dr Fatih Birol
Executive Director
International Energy Agency

About the Author

Neil Hirst is the Senior Policy Fellow for Energy and Mitigation at the Grantham Institute, Imperial College London. After occupying a variety of posts for the UK government, he was the Director for Technology and later the Director for Global Dialogue at the International Energy Agency.

In a wide-ranging career in international energy policy, Mr Hirst has been the Energy Counsellor at the British Embassy in Washington, and Chairman of the G8 Nuclear Safety Working Group. He has worked on energy finance on secondment to Goldman Sachs in New York and he has undertaken consultancy for the World Bank.

At the IEA, he led a programme on climate mitigation work in support of the G8. He pioneered the IEA's flagship technology publication Energy Technology Perspectives, and he forged closer links with China, India and Russia. Currently he is working with the Chinese government on international energy governance.

Mr Hirst holds a first class degree in Politics, Philosophy and Economics from Oxford and an MBA from Cornell.

Contents

Chapter 1

Introduction

1. Thinking About Energy

Energy policy is about the kind of society that we want to live in and the effort and cost that we are willing to expend to achieve it. It is about poverty alleviation and economic growth, human health, and welfare. It is about the environment, including climate change, and pollution, and it is about maintaining the reliable energy services that we all, in advanced societies, depend upon every minute of the day and night. It is in part technical, but at its heart, there is political and moral choice.

A good way to start thinking about energy is to get on an exercise bike. After all, before the industrial revolution, the human body, admittedly with some help from animals and the occasional wind and water mill, was the main source of energy available to humanity. Most people can just about sustain 130 Watts for half an hour, enough to power two old-fashioned incandescent light bulbs. Power stations in developed nations have the capacity to generate more than 15 times that continuously for every man, woman, and child. On top of this electric power, we use large amounts of energy for transport, heating and cooling buildings, construction, and manufacturing. In advanced countries, we each consume energy that is equivalent to more than 4 tonnes of oil each year. We are not just dependent on energy, we are dependent on prodigious quantities of energy.

1

Energy is big business, indeed by some measures the biggest business of all. One may think of the major international oil companies such as EXXON and BP, but today, most oil and gas production is undertaken by state companies belonging to oil producing countries and to the major developing nations who are now amongst the biggest users of energy. Saudi Arabia's oil company, Aramco, is expected to become the world's most valuable company, when, as planned, its shares become available in stock markets.[i]

Energy is a major player in world trade and investment. Governments of exporting countries depend on energy revenues and many people, not all of them wealthy, depend on the dividends of energy companies. The struggle for access to energy resources around the world is a continuing source of geopolitical tension.

The historic contribution of the energy industries to raising living standards around the world has been extraordinary, especially in the last few decades. But more than 700 million people still live in extreme poverty. A great many others are far from enjoying the quality of life that most readers of this book enjoy, and they will need a lot more energy services to meet their aspirations. How this feeds through into total energy demand will depend, in part, on the progress that we can make in improving energy efficiency. We now realise that energy is also the main source of the emissions of greenhouse gases that are causing irreparable damage to the atmosphere. We have to find a way to sharply reduce these emissions while, at the same time, continuing to increase the services that energy provides.

Even with the most rapid switch to low carbon energy that seems feasible, we will still be reliant upon the fossil fuels — coal, oil, and gas — for several decades to come. The economies of importing and exporting countries all depend on international energy markets. The extraordinary volatility of energy prices, while favouring consumers at the time of writing, is a headache for importing and exporting nations alike and makes it difficult to plan energy investments of any kind.

---

[i] Bloomberg, *Too Big to Value: Why Saudi Aramco is in a League of its Own*, 7 January 2016.

2. What is Energy Policy?

In taking energy policy decisions, each government has to balance its own priorities for security of supply, energy costs, the local and global environment, economic growth and development, jobs, poverty eradication, import dependency, resource income, technology leadership, and peaceful international relations.

These challenges look quite different in different parts of the world. A developed country with a cold climate and abundant fossil fuel or hydro-electricity faces very different energy challenges from those of a developing country with a hot climate and without natural resources. There are also cultural differences. Nuclear power is acceptable in some parts of the world but not in others.

Nevertheless, the main energy challenges are essentially global. Energy markets are global, we have one atmosphere to protect, and, increasingly, energy technologies and the businesses that deploy them are global. Even the trends in government decision-making are global.

Some of the main elements of sound energy policy are well known. We must improve the efficiency with which we produce and use energy. We must invest for the future, especially in less polluting energy supply. We must conduct research to enhance the available technologies. The developed nations must support the developing nations. We must keep costs down for taxpayers and energy consumers. We need international agreement to keep energy markets open and manage emergencies.

But in practice, there are barriers and there are conflicts. That is the conundrum. As a result, we are definitely not on course today to meet either the social or the environmental objectives of energy policy. This book seeks to explain why we are not on course and the options available for a clean, affordable, and secure energy future.

Energy is a highly charged topic, as it should be. Important moral and political issues are at stake. In this book, I have tried to describe the facts and the issues in an objective way so that the reader can make up his or her own mind. Only at the end, in the conclusions in Chapter 11, do I give my recommendations on how we can solve the energy conundrum.

People arrive at the energy scene from different perspectives. Some may have a general interest in the subject. Others may come at it from a particular angle. Someone pursuing a career in oil exploration might feel that they have little need to know about energy and climate change. Someone working for an environmental NGO might feel that the achievements of today's conventional energy industries are irrelevant to them. But they would be wrong. The carbon-reducing transition that is already in progress is set to have profound implications for the fossil energy industries. Equally, fossil fuels will be with us for some time yet, and those concerned for the environment have to address themselves to the practical problems of minimising emissions during a transition period. A broad understanding is needed, and that is what this book intends to provide.

The energy industries themselves are highly diverse. In some ways, it is misleading to speak of energy technology because there is nothing in common between the different technologies that extract energy from fossil fuels, nuclear fission or fusion, falling water, hot underground strata, the wind, the sun, or the ocean, all of which are significant sources or potential sources. However, one thing that many areas of energy technology do have in common is that they are at the forefront of technological progress in their respective fields. The technological achievements of the energy sector have been astonishing and there is every reason to believe that they will continue to astonish in the future.

Is there such a thing as energy policy? Or are there just a collection of quite separate political, economic, and ethical issues that happen to be in the energy sector? All our major energy challenges are connected in complex ways both globally and nationally. Energy security, energy affordability, and the protection of the environment, the three pillars of energy policy, are inextricably linked. It is no good coming up with environmentally acceptable energy options that cannot provide secure affordable energy and support economic development. But, equally, we must not follow a pathway that is economically attractive in the short term but, ultimately, environmentally disastrous.

3. Themes of this Book

A major theme of this book is that there is such a thing as energy policy and that we must understand its complexities and interactions if we want to find workable solutions in the energy field. Economics, diplomacy, environmental studies, and technology are all highly relevant, as are a variety of social disciplines. But none of them has the full picture. Too much of the literature, and perhaps of policy-making, approaches energy policy as though it was a branch of one of these fields.

In this brief introduction, I have illustrated the all-pervading role of energy in modern society, the fact that our energy challenges are global and interrelated, that different countries view these challenges from different perspectives, and that policies have to be found that successfully resolve the tensions between different objectives. We need an integrated approach to the energy conundrum that we face.

The book seeks to elucidate the issues of energy policy in a narrative style. This contrasts with the mathematical projections into the future, known as scenarios, presented in many energy reports. These scenarios definitely have their uses. I have worked on scenario publications at the International Energy Agency (IEA) and references will be made to some of the most recognised projections at many points in this book. They impose a degree of rigour in thinking about the future because the numbers have to add up and the assumptions used in the models have to be specific. If well presented, they offer very clear pictures of what the future might look like, and many policymakers find that helpful. However, the user, inevitably, will only have a limited understanding of the assumptions that have been fed in, many of them on very uncertain matters. There is a risk of obscuring, rather than elucidating, the underlying issues which, I hope, has been avoided here.

The book starts with a description of the development of the fossil fuel industries that provide most of our energy today, the geopolitical tensions that arise from the rising economic strength and energy demand of the East, and the problems of securing supplies and managing emergencies. These problems are not new, but their nature is changing. Then we move on to the essential topic of energy and human welfare, setting out the immense gains that have been made, and also the immense needs that still exist, especially in Asia and Africa. That is what energy is for.

The next topic is climate change, one of the greatest threats facing humanity, and the international effort to curb carbon dioxide emissions. We are making progress towards a cleaner energy world, but definitely not fast enough. Next is a review of the technological options which provide our best hope of speeding this up. How soon will advances in renewables and other areas enable us to switch to clean, environmentally friendly sources?

Then we turn to the topics of energy finance, the reduction of fossil fuel subsidies, and carbon pricing. There is then a chapter on energy efficiency followed by a chapter on energy markets. These are all critical topics for the efficient implementation of a low carbon strategy.

The last topic is global energy governance. Solving our energy problems will require global leadership at the highest level. Do we have the right international energy institutions to make this possible? Finally, in the last chapter, I give my own, more subjective views, on how we can best solve the energy conundrum.

There follow two case studies in which readers are invited to try some energy policy making of their own. You are asked to come up with solutions to the difficult policy dilemmas faced by the energy ministers, respectively, of two imaginary countries, one rich and the other poor. You are encouraged to give these a try. They are thought-provoking. The case studies are also suitable for the use of classes of graduate students, working in groups. There are no "correct" solutions. Much of our energy future will depend on social attitudes, so the choices that today's students make may provide significant clues as to the future direction of travel.

4. Sources and Acknowledgements

Major sources are the IEA's two annual publications, World Energy Outlook (WEO) and Energy Technology Perspectives (ETP). The various reports of the US Government's Energy Information Administration (EIA) and BP's annual Energy Outlook and Statistical Review of World Energy have also been used, together with Greenpeace's Energy Revolution, also published annually. The reports of Working Group III of the Intergovernmental Panel on Climate Change (IPCC), the global expert group on climate change, are another important source. I also recommend the annual Human Development Reports of the United Nations Development Programme (UNDP). These are all valuable publications for those wanting to deepen their studies.

I would like to thank my wife, Caroline for suggesting the book's title and helping with the overall structure, as well as for her patience. I am also grateful to those who have generously given their time to reading drafts and offering helpful comments, colleagues at the Grantham Institute, Ajay Gambhir and Sheridan Few, former colleagues and leading international energy experts Joan Macnaughton CB and Graham White CBE, and friends and relatives who are not energy specialists who offered comments from the perspective of discriminating general readers, my brother Prof Ian Hirst, cousin David Hirst, Roger Hird, John Neve, and Tim Yarker.

I would also like to thank the Grantham Institute, Imperial College, where I work, for making it possible for me to write the book. I am especially grateful to Jane Sayers, Senior Commissioning Editor at World Scientific Publishing, for editing the draft and for her constant encouragement and also to Sree Meenakshi Sajani and Jennifer Brough who prepared the final print and managed the book through to publication. Responsibility for the final text is my own.

Chapter 2

History and Geopolitics of Fossil Energy

1. Introduction

Coal was and to some extent remains the bedrock of industrialisation. Oil is the most traded commodity and a major source of geopolitical tensions. Gas has been a national or regional fuel, but now that gas is increasingly traded in liquefied form, it is beginning to take on some of the same geopolitical significance as oil. These industries have shaped the world and will continue to do so, at least for several decades to come. We cannot make wise energy policy choices for the future without understanding this process. This chapter outlines some of the most salient events in the history of these energy industries, their impact on the world that we live in today, and the questions that they pose for the future.

One might ask, why do we need to study fossil fuels? Surely, the future is with low carbon and renewable forms of energy? They will be clean and locally produced and they will bring to an end the horrible pollution and world political tensions that have come with our addiction to coal, oil, and gas. Yes, this is definitely the direction in which we need to go. The International Energy Agency (IEA) has described how, with sufficiently strong political will, we can decarbonise the world's energy economy fast enough to limit global warming to 2°C, the target that has been set by world leaders. But they say that the pathway is very tough.[i]

We need to press on with this transition as fast as we can. But this is going to take time, even assuming that it goes as smoothly as we may hope. The reasons for this are discussed in Chapter 5. Partly, it is the result of the sheer inertia of the vast global energy systems. Although the costs of renewables are declining, they are still above those of fossil fuels in many uses and locations. We are trying to develop more flexible electric systems, but at present variable renewables still require backup, often in the form of fossil power. Electricity from renewables has still not penetrated significantly some of the largest energy markets, for instance, for road vehicles, industry, and the heating of buildings. According to the IEA, even if we succeed in meeting the 2°C target, we will still be relying on fossil fuel for 61% of our energy as far ahead as 2040.[ii] The pathway that we are on today, even taking account of the decarbonisation efforts that governments proposed at the Paris climate change summit of November 2015, we will still be 74% dependent on fossil fuels in 2040.

So fossil fuels are still important. We will depend on them for our vital needs and for economic progress around the world for several decades to come, and the level of carbon emissions during the transition period will depend on the choice of fossil fuels and the efficiency with which they are used.

2. The Industrial Revolution

In Christian mythology, the first man was condemned to a life of physical toil because of his disobedience to god. "In the sweat of thy face shalt thou eat bread".[iii] And indeed from ancient times, mankind has relied largely on his own bodily energy, with some help from domesticated animals, water and wind power, for his survival.

That changed with the coming of the Industrial Revolution in Britain, generally attributed to the years 1780–1850. A key feature of this revolution was the conversion of energy from coal into mechanical power,

---

[i] International Energy Agency, *World Energy Outlook*, 2016.
[ii] International Energy Agency, *World Energy Outlook*, 2017.
[iii] *King James Bible*, Genesis Chapter 3 v. 19.

thereby vastly increasing the power at humanity's disposal. The output of coal in Britain increased three and a half times between 1775 and 1830. By 1850, pig iron output had risen by 35 times and by 1873 British output was equal to that of Europe and the United States combined. Britain had become the "workshop of the world".[iv] The steam railway revolutionised transport. More than 3,000 miles of track were laid between 1830 and 1846, a breath-taking achievement for the times. During this period Britain, a small island with less than 2% of the world's population, became the dominant world power.

The true nature of the industrial revolution has been the subject of extensive scholarly debate, no doubt because it was such a seminal process. If we can understand the industrial revolution, perhaps this will give some insight into the industrialised world in which we now live. The traditional view highlights the contribution of specific technological advances, such as steam power, originally for pumping water from mines and then for railways and industrial processes, especially for cotton material manufacture.

Some have questioned whether there was a revolution at all and whether the British economy of the time can be called industrial. They point to the fact that some of the most significant changes in the British economy had taken place in the first half of the 18th century and before and that these were based on developments in commerce, including the disgraceful slave trade, finance and agriculture, rather than manufacturing. These commentators put more emphasis on the social, economic, and political factors that made industrial enterprise profitable and attractive.[v] According to Arnold Toynbee, "The essence of the industrial revolution is the substitute of competition for the mediaeval regulations which previously controlled the production and distribution of wealth — Europe owes to it the growth of two great systems of thought, economic science, and its antithesis, socialism".[vi] It is an interesting question whether the

---

[iv] Hilton, B. *A Mad, Bad and Dangerous People? England 1783–1846*, New Oxford History of England, Clarendon Press, 2006.

[v] Allen, R. *The British Industrial Revolution in Global Perspective*, Cambridge University Press, 2009.

[vi] Toynbee, A. *Lectures on the Industrial Revolution in England*, 1884. Reprinted by Newton Abbott 1969.

development of a more competitive and efficient economy might have taken other directions. However, it is incontrovertible that the industrialisation of Britain in the 18th and 19th centuries was built on coal.

The industrial revolution raised, to a remarkable degree, and from the first, many of the same issues about the nature of our industrial society and its use of energy that are the subjects of such profound debates today. Railways had to win the approval of parliament and, in doing so, to overcome the vested interests of the owners of the land required, and of the canals and toll roads with which they competed. There was vigorous lobbying and propaganda on both sides. All kinds of reasons were offered for opposing the new lines, some of them bizarre to modern understanding but others quite reasonable at the time. The early accident rate was terrible. The poet Wordsworth led the opposition to the "contamination" of the Lake District by a railway, which was subsequently built.[vii] The introduction of mechanical weaving machinery was socially disruptive at the time and was violently opposed by the handloom weavers whose jobs were jeopardised.

Not everyone thought that the industrial revolution was a good thing at the time. It was far from clear that it brought any early improvement to the conditions of life of ordinary people, and there is nothing new in today's concerns about the environmental consequences. The novelist Dickens, famously, was a savage critic of the social and physical conditions of workers and of industrial pollution. Here is his description of Coketown in his novel Hard Times, published in 1854:

> It was a town of red brick, or of brick that would have been red if the smoke and ashes had allowed it; but as matters stood, it was a town of unnatural red and black like the painted face of a savage. It was a town of machinery and tall chimneys, out of which interminable serpents of smoke trailed themselves for ever and ever, and never got uncoiled. It had a black canal in it, and a river that ran purple with ill-smelling dye, and vast piles of building full of windows where there was a rattling and a trembling all day long, and where the piston of the steam-engine

---

[vii] Leigh, D. and Lendis, J. *Motion and Means: Mapping Opposition to Railways in Victorian Britain*, 1999.

worked monotonously up and down, like the head of an elephant in a state of melancholy madness. It contained several large streets all very like one another, and many small streets still more like one another, inhabited by people equally like one another, who all went in and out at the same hours, with the same sound upon the same pavements, to do the same work, and to whom every day was the same as yesterday and to-morrow, and every year the counterpart of the last and the next.[viii]

We now know that, in addition to the very obvious local pollution that Dickens described so eloquently, coal is also the worst of all the fossil fuels as an emitter of greenhouse gases. In fact, coal produces about twice as much CO_2 as gas, which is the next largest source of electricity, for the same amount of power.

Friedrich Engels was also a severe critic of the social conditions of the industrial revolution. His 1848 publication, "The Conditions of the Working Class in England", was the precursor to the Communist Manifesto which he wrote with Karl Marx. Social and environmental legislation, together with more representative government, have since moderated the darker aspects of the industrial revolution in the UK and elsewhere. But communist ideology went on to be influential in many parts of the world.

Profound questions are still being raised as to whether following the Western path of industrialisation is the right way forward for humankind.[ix] Nevertheless, it is due to industrialisation that, in the memorable words of Martin Wolf in his 2013 lecture at Imperial College,[x] "the majority of people in the Western World today enjoy a living standard undreamed of by princes in the pre-industrial period". We are not going to give that up voluntarily.

Only about one fifth of the world's population live in these rich industrialised nations.[xi] The other four fifths of mankind understandably aspire to the same standard of life and, in many cases, are well on their way to achieving it. They are not going to give up their aspirations either.

[viii] Charles Dickens, *Hard Times*, 1854.

[ix] Jackson, T. *Prosperity Without Growth*. Earthscan/Routledge, 2009.

[x] Personal recollection.

[xi] i.e. the OECD countries.

The West cannot pull up the ladder whereby it ascended. Nevertheless, the spread of industrialisation to the majority of mankind is making huge new claims on natural resources and putting us on a path that spells disaster for the environment.

3. Coal

Although she became the world's main coal supplier during the industrial revolution, Britain has, in fact, only a minute fraction of the world's coal reserves. Nature has distributed these liberally around the world, in many cases to where they are most needed. The countries with the largest coal deposits are the US, Russia, China, India, and Germany, in that order.[xii] But many other countries, such as South Africa, Poland, Colombia, Indonesia, also have substantial deposits.

Where Britain led, the rest of the world has followed. More than 150 years after the first industrial revolution, coal remains, in large measure, the basis for global industrialisation. In many rapidly developing countries, coal is the dominant fuel for electric power. For them, coal is the cheapest, most accessible, and most secure form of energy and being produced at home, it does not drain foreign currency reserves. China's extraordinary economic progress in recent decades has been almost entirely fuelled by a prodigious increase in the production of coal, which, in 2016, still provided 67% of China's electric power. India is 76% coal-dependent. Indeed, coal is still the largest source of electric power in the world as a whole, contributing 37%.

Perhaps what is most extraordinary is that still, in 2016, more than 20 years since the climate treaty in which the developed nations committed themselves to taking the lead on carbon reduction, 31% of electricity in the US and 23% in Europe was still coal-powered.[xiii] The overall figures are staggering. In 1850, Britain produced about 50 million tonnes of coal.[xiv] Today the world produces more than 5 *billion* tonnes of coal, of

---

[xii] US Energy Information Administration (EIA) Statistics.

[xiii] International Energy Agency, *World Energy Outlook*, 2017.

[xiv] Polard, S. *A New Estimate of British Coal Production 1750–1850*, The Economic History Review New Series, Vol. 33 No. 2. 1980.

which about half is produced and consumed in China.[xv] As a result, coal combustion is the world's largest source of CO_2 emissions.

According to the IEA, after prodigious growth between 2000 and 2010, world coal demand fell slightly in 2015, for the first time since the late 1990s. The future is very uncertain, but the IEA are projecting that demand will be static to 2020 but that, thereafter, increased use in India and South East Asia will just outweigh declines in Europe, the US, and China.[xvi]

If we are going to protect future generations from global warming we must find a way to moderate and then reduce world use of coal, or perhaps find ways to reduce its environmental consequences. As I will discuss later, there are signs that this is beginning to happen, but not fast enough. Besides being the largest source of greenhouse gas emissions, coal is also the worst offender in terms of local atmospheric pollution. The technologies already exist for removing most local atmospheric pollutants, such as sulphur dioxide and nitrogen oxides, from the emissions of large power stations. They come at a cost, but they are increasingly widely used.

Nor are the coal reserves likely to run out any time soon. According to the IEA, "the resource base is easily sufficient to meet any plausible level of demand for decades to come".[xvii] Even though developed nations are cutting back on the use of coal, world demand is static or rising slightly because of the industrialisation of developing nations. We need to more than halve world coal consumption by 2040 to contain climate change to 2°C. But on IEA projections coal consumption may, by then, be slightly larger than it is today.[xviii]

The cost of coal production varies, depending in part on whether the coal is taken from the surface or from deep mines. As a general rule the cheapest coal comes from regions, such as Wyoming in the US, where it lies close to the surface and, once the "overburden" of earth has been removed, it can be excavated with giant shovels. Deep mining, which requires the sinking of shafts and underground tunnelling is usually more

[xv] International Energy Agency, *World Energy Outlook*, 2017.
[xvi] *Ibid.*
[xvii] *Ibid.*
[xviii] *Ibid.*

costly. Also, coal is expensive to transport, especially over long distances by rail. For these reasons, the cost of coal, and its competitiveness with alternative energy sources, varies substantially from place to place. In comparison to oil and gas, coal production has relatively low capital costs but relatively high "variable" costs incurred in production. This gives coal a somewhat greater degree of price stability, at least in the medium term. In recent years, international coal prices have been depressed, along with the prices of other fossil fuels and most commodities. This largely reflected the slowdown of the global economy and especially the Chinese economy, and competition from cheap gas, especially in the US. However, as of May 2017, there has been some recent recovery as China has begun to reduce its production.

Unlike oil, most coal, about 85%, is consumed in the country where it is produced. Nevertheless, international trade in coal is substantial. Europe, Japan, China, and India are the biggest importers and Australia and Indonesia are the main exporters.

Coal consumption in Europe and the US is now declining. Possibly, the peak of coal consumption has also, now, been reached in China. But, as already mentioned, coal demand is still growing in India and in much of South East Asia. It may also grow in Africa. An absolutely central question for the future of the climate is whether, in these rapidly growing regions, the switch to lower carbon alternatives can be achieved in a way that is positive for, and does not at all delay, economic growth and poverty reduction.

As far as international trade in coal is concerned, much will depend on China's future policies. Most of China's coal is deep mined. The most accessible deposits have already been mined, and the remaining deposits tend to be more difficult to mine or located in the West of China, especially Xinjiang province, many thousands of miles from the main centres of demand. Coal from Australia or Indonesia has the potential to compete strongly in Chinese markets, especially as Chinese labour costs rise. Chinese coal demand is so huge that any significant shift towards imports would have a big impact on international coal markets.

In 2010, I was asked, as part of a three person panel, to advise the World Bank on whether to invest in a huge coal power station at Medupi in South Africa. We faced a serious dilemma. South Africa is a major supplier of electricity to 11 other countries in sub-Saharan Africa. South Africa faced an immediate shortage of electric power that had already crippled its economy. The Medupi plant, which consisted of a complex of six large (800 MW) power stations would emit more than half a million tonnes of CO_2 a year, undoubtedly a major climate change hazard. Yet it also had the potential to increase South Africa's power supply by 12%.

Of course the ideal solution would have been to work on South Africa's poor level of energy efficiency and to introduce more renewables, especially solar and wind power. Indeed, the World Bank package included some investments of this kind. But we did not believe that these alternatives, on their own, could provide the large and immediate increase in power that was so vitally needed. Even in the most advanced economies, the transition to low carbon alternatives has been gradual.

We also considered that CO_2 emissions per head in Africa are less than a quarter of the world average and less than a 10th of those in the West. South Africa's own emissions per person are higher, reflecting a concentration of mining industries, but they are still less than half those of resource based economies in the developed world, such as Canada or Australia.

At the Copenhagen climate summit in 2009, world leaders agreed that, for developing nations, "economic and social development and poverty eradication are the first and overriding priorities". It's an obvious conclusion. In an ideal world there would be no conflict between the social objectives of energy policy and the need to reduce carbon emissions. Unfortunately this is not always the case. It was a tough decision. We concluded that the World Bank should support Medupi but that the Bank should also commit itself to supporting a clean energy transition for South Africa and a plan to assist South Africa to achieve long-term greenhouse gas savings comparable to the Medupi emissions.[xix]

Did we reach the wrong conclusion? Since then the World Bank has announced that it will restrict funding for new coal plants in developing

---

[xix] Davidson, O. Hirst, N. and Moomaw W. *Recommendations to the World Bank Group on Lending to South Africa for ESKON Investment Support Project that includes a Large Coal Burning Power Station at Medupi*, World Bank, 2010.

countries, except in "rare circumstances". This is widely taken as an effective ban. The President of the bank, Jim Yom Kim said that, "the world's top priority must be to get finance flowing and get prices right on all aspects of energy costs to support low carbon growth".[xx] This is not quite the same thing as the Copenhagen formula, which gives clear priority to social objectives. Is it right to ban new coal outright in developing nations? Different views are possible. This will be discussed further in the concluding chapter.

Coal mining, with its large labour force has, of course, had a huge social and political impact in many parts of the world. There have been disputes about the pay and conditions of mineworkers and continuous efforts of mining unions to improve them and to protect mining jobs. Anyone in Britain who lived through Mrs Thatcher's governments of the 1970s will know all about the history of strife in the coal industry during that period. Governments have fallen and some of the greatest threats to the UK's security of energy supply and social cohesion in the 20th century have come from mining disputes. More generally, wherever coal mining is a major industry, policies for running it down rapidly will face tough social and political challenges.

As long as coal remains a major energy source, the options for improving the efficiency of coal use will be important for moderating emissions. Coal varies greatly in quality and there are significant gains to be made from using higher grade resources and also from washing coal to clean out impurities. The technology of coal power stations is discussed further in Chapter 6. Generally speaking, efficiency is enhanced by raising the temperature and pressure of the boiler. In "super-critical" boilers, water explodes into steam without forming bubbles. The latest and most efficient plant rejoices in the title of "ultra-super-critical". There are, of course, alternatives, available today, to the use of coal in power stations, including nuclear power and renewables. Even switching to gas offers a major environmental advantage provided that emissions in gas production and transmission are well controlled. Switching away from coal in power

---

[xx] Plumer, B. 17 July 2013. The *World Cuts off Funding for Coal. How Big an Impact will that Have? Washington Post* wonkblog. Available at http://www.washingtonpost.com/blogs/wonkblog/wp/2013/07/17/the-world-bank-cuts-off-funding-for-coal-how-much-impact-will-that-have/. Accessed 2 July 2014.

generation will be a vital step towards decarbonising the global economy. How this is to be achieved, the technologies involved, and their availability and cost is a topic for Chapter 6.

But it is important to remember that power generation, although it represents the largest single use of coal worldwide, is far from being the only use. About 40% of coal is used elsewhere, mainly for industrial use, including specialised metallurgical coals for steel making. Generally speaking, replacing coal in these uses is going to be more difficult. There is expected to be a continuing demand for metallurgical and other industrial uses where there is no readily available alternative.

Is coal a dying industry? Not necessarily. It is true that coal is in decline in the US and Europe as a source of electric power. In the US this is mainly the result of cheap shale gas, which competes directly in the power station market. It may also be threatened by environmental regulation. New carbon emissions regulations being brought in by the US Environment Protection Agency under President Obama, already held up by a decision of the Supreme Court, have now, in June 2017, been struck down by President Trump. But a number of US states and cities continue to pursue low carbon policies. Peabody Energy, the world's largest private sector coal producer, which filed for Chapter 11 bankruptcy protection in April 2016 has now emerged, a year later. But the outlook for coal in the US remains poor.

In Europe, coal is declining due to low carbon government policies and environmental regulation including the Emissions Trading System (ETS). However, the penalty for carbon dioxide emissions under the ETS is too low, for the time being, to have a dramatic impact on coal demand and the decision to phase out nuclear power in Germany is leading to a revival of coal. Hopefully, that will only be temporary.

In China, coal is now probably in gradual decline, but this comes after a decade of stupendous growth. In India and South East Asia, and probably in Africa, coal demand may continue to grow, depending on the accessibility of cost-effective alternatives.

The outlook for coal in India is especially important. The Indian Ministry of Power announced in December 2016 that no additional coal capacity was needed to be commissioned for the period 2017–2022.[xxi]

[xxi] Central Electricity Authority, Ministry of Power, Government of India, *Draft National Electricity Plan (Volume 1) Generation*, December 2016.

This partly reflected the growth in solar and wind energy, but it also reflected the large amount of coal capacity that is already under construction. In principle, Indian government remains committed to efficient "super-critical" coal generation as part of its low carbon growth strategy.

Coal remains vital, today, to many economies, especially in the developing world. But its position is under threat from gas, especially shale gas, from environmental and climate policy, and from renewables. To meet climate objectives coal must either be phased out or its emissions must be stored. The phasing out of coal in the developing world must be done in a way that is fair and does not deprive the population of development opportunities. That is likely to require substantial international support, a topic discussed further in Chapter 4. During the transition stage, efficiency improvements in coal power stations can make an important contribution to reducing emissions.

The longer-term future of coal in a low carbon energy world will depend on pumping the carbon dioxide emissions from coal combustion underground, a technology known as Carbon Capture and Storage (CCS). CCS has made a slow start because, in the absence of carbon pricing, there is no business case and governments have preferred to invest in renewables. It makes coal plants more costly as well as reducing their efficiency. But many analysts believe that widespread adoption will be essential to meet the very low levels of carbon emissions required by climate targets. The outlook for this technology may change as carbon restrictions tighten and energy prices rise. In the meanwhile, it would make sense for the coal industry, together with the gas industry which also stands to benefit, to work closely with governments to bring forward a technology that could be vital for their future. We will return to the prospects of CCS in Chapter 6, on technology.

4. Oil

The politics of coal are mainly domestic. It is oil, which first became important as an international commodity in the late 19th century, that has placed energy at the centre of geopolitics.

To a large degree, in modern societies, transport requires oil. We depend on it totally for many essential services and for our quality of life.

The catastrophic consequences of running out of oil, as well as the management of oil supply crises are discussed further in the next chapter.

We use an awful lot of oil. In advanced industrial societies we each use, directly or indirectly, about one and a half tonnes of oil each year. That is about 10 barrels or 1,600 litres.[xxii] Developing countries are generally even more dependent on oil than advanced societies because they are still building their infrastructure such as power stations and railways. Goods are often transported by heavy lorries on poor roads. Small diesel generators are in common use. These are inefficient, costly, oil-intensive options. That is why developing nations, with their limited reserves of foreign currency, are so vulnerable to oil price fluctuations.

It is hard to over-state the economic and political impact of the international oil industry. Of the 10 largest corporations in the world, by sales, listed in the Fortune Global 500 for 2016, four were oil companies and two of the others were car manufacturers.[xxiii] The total revenues of the four oil companies were about $1 trillion, and this was in a year of depressed oil prices. The Saudi Government has said that it intends to float some of the shares in Saudi Aramco in the stock market. As already mentioned, assuming that this happens, the share price is expected to rate Aramco as the most valuable quoted company in the world.

Unlike coal, oil has not generally been distributed by nature where it is most needed. Or at least that is true of the conventional oil fields which have been the main source of production so far. The revolutionary possibilities of opening up non-conventional sources will be discussed later. Nearly, half of all the conventional reserves are in the Middle East. Africa and South America are also major sources. Saudi Arabia, Iran, Iraq, Kuwait, United Arab Emirates, Russia, Libya, and Nigeria are amongst the best endowed nations.

The major industrialised nations, including Europe, the US, and Japan have about 14% of the world's conventional reserves between them. As a result, with the partial exception of the US, they have historically depended on imports. It is this geographical mismatch between demand

---

[xxii] International Energy Agency, *Key World Energy Statistics*, 2016.
[xxiii] Fortune Magazine, July 2013.

and supply, combined with the fact that, as a liquid, oil is relatively easy to transport, that has driven the globalisation of the oil industry.

The world oil market is perhaps the most striking exemplar of Adam Smith's dictum, "It is not from the benevolence of the butcher, the brewer, or the baker that we expect our dinner, but from their regard to their own interest".[xxiv] There have been many conflicts, both political and military, between the oil producers and consumers of the world but, by and large, trade has continued to flow. Keeping international oil markets functioning is one of the most important requirements for international peace and prosperity.

This section will describe the geopolitical impact of the oil industry over the past century, and the related economic impact of oil price volatility. It goes on to discuss the factors that are changing the face of global oil today and the outlook for supply and demand.

Access to international oil supplies, and the control of supply routes, have been strategic, political, and sometimes military objectives of the major powers over the past century. Oil has constantly interacted with global politics. From the period leading up to the First World War when Winston Churchill decided to convert the British fleet from coal to oil,[xxv] access to oil reserves, primarily in the Middle East, became a dimension of British diplomacy. This started with the creation of the Anglo Persian oil Company, the parent of BP, and led to the sponsorship of the regime of the Shah of Iran until its collapse in 1979.

During the Second World War, the need to gain access to oil reserves in the Caspian region was one of Hitler's reasons for launching an unprovoked attack on the Soviet Union, disastrously for him as it turned out.

During the 1930s, Japan was seeking to establish its status as a great power by imperial aggression, originally in China and then in Indo-china. However, the Japanese economy and war machine were critically dependent on imports of oil from the US. In 1941, Franklin D. Roosevelt imposed an oil embargo on Japan. This placed the Japanese in a position where they had to choose between "war and subservience". Only by capturing the Dutch East Indies could the Japanese find substitute oil supplies.

[xxiv] Smith, A. *An Inquiry into the Nature and Causes of the Wealth of Nations*, 1776.
[xxv] Jeffrey, R. *The Politics and Security of the Gulf*. Routledge, 2010.

They knew that this would mean war with the US. Although the overall causes of war with America were complex, this was the proximate cause of Japan's attack on Pearl Harbor. "The attack on Pearl Harbor was essentially a flanking raid in support of the main event, which was the conquest of Malaya, Singapore, the Indies and the Philippines".[xxvi] The lesson is not lost on government strategists today.

During the Arab–Israeli war of 1973, when the US gave its support to Israel, the Arab members of the Organization of the Petroleum Exporting Countries (OPEC) imposed an oil embargo on the US. There was a crisis of oil supply in the US, with lines at the petrol stations. World oil prices quadrupled very quickly. The producers were using their market power for the first time. These events represented a shock to Western interests that appeared catastrophic at the time and caused a huge transfer of wealth and power to the OPEC producing nations. There were divisions amongst Western governments and this "was widely regarded as the worst crisis, and the most fractious, to afflict the Western alliance since its foundation after World War II."[xxvii] OPEC had found its strength and further price increases continued into the early 1980s.

According to Anthony Sampson,[xxviii] "The (US) Defence Secretary, James Schlesinger, began hinting that the embargo might lead the government to use force in the Middle East; and hints continued to be dropped from the Pentagon". The fact that the US decided, at this moment not to intervene militarily to secure its oil supplies can be regarded as a turning point of history. In fact, the response that was championed by Secretary of State Henry Kissinger was not military but took the form of a defensive alliance amongst oil consumers. This was the IEA, a club through which the advanced consuming nations committed themselves to a joint oil security programme. They agreed to build oil stocks, to a common emergency response mechanism, and to longer-term measures aimed at reducing oil import dependency. Since then, the IEA has evolved to address the full

[xxvi] Record, J. *Japan's Decision to go to War in 1941: Some Enduring Lessons.* US Army War College, Strategic Studies Institute. February 2009. Available at http://www.strategic-studiesinstitute.army.mil/pubs/summary.cfm?q=905. Accessed 2 July 2014.
[xxvii] Yergin, D. *The Quest.* Allen Lane (Penguin Books), 2011.
[xxviii] Sampson, A. *The Seven Sisters.* Hodder and Stoughton, 1975.

range of international energy issues and has become the world's leading international energy body. However, its membership is still limited to the developed nations of the OECD. The important role that the IEA plays in today's international energy governance, and the progress of the IEA towards becoming a more broadly based organisation, are discussed in Chapter 10.

There have, of course, been innumerable other instances of oil being used as a political weapon. These include embargoes imposed on oil supplies to Zimbabwe, then known as Rhodesia, South Africa, and the embargo on sales of oil by Iran that is now in the process of being lifted. Iraq's invasion of Kuwait in 1990 was the most direct deployment of military power to capture oil assets in recent history. The success of the current struggle to defeat the extremist Islamic organisation, the Islamic State of Iraq and the Levant (ISIL) depends, to a considerable degree, on cutting off its oil revenues.

It would be naïve to think that this geostrategic role of oil is a thing of the past. The oil market continues to suffer from instabilities and tensions and the international struggle for access and control continues. This topic will be discussed later.

Besides these geopolitical impacts, the volatility of oil prices has had, and continues to have, an important impact on the global economy. Most of the costs of conventional oil supply are in exploration and development. Once a field has been developed, the marginal costs of producing the oil are fairly low. This means that in the short term, production may continue even if there has been a big fall in price.

Similarly, oil consumers are typically motorists or trucking operators who have already invested in their vehicles and will not greatly vary their travel habits in response to rising oil prices. In economists' terms, the short-term price elasticity of both supply and demand is low. This creates the possibility of wild swings in which the price of oil may depart substantially from its marginal production cost for extended periods. Associated with these price swings is an investment cycle. High prices lead to an investment boom and, in due course, overproduction and a price collapse. Low prices lead to the reverse. The problem is common, to varying degrees, to internationally traded commodities as a whole but oil has the biggest economic impact.

The high volatility of oil prices has been a problem for producers and consumers alike, though of course they have different views on where the price should be struck. Not surprisingly, the history of oil has been a story of successive efforts by industry and by governments to control the price through some kind of a cartel.[xxix]

In 1883, in the early days of oil production in the US, John D Rockefeller successfully merged a wide range of oil interests into the Standard Oil Trust, which at one time controlled up to 90% of the US industry and dominated the global market. This was the most notorious of the monopolistic American trusts of the period and it was a prime target of President Theodore Roosevelt's "trust busting" campaign through the Sherman Antitrust Act. In 1911, the US Supreme Court took a key decision that lead to the break-up of the Standard oil Trust into 34 separate companies.[xxx]

This was definitely not the last of the efforts to cartelise oil production and control prices. In 1928, the leaders of the "big three" international oil companies of the day, Standard Oil of New Jersey, BP, and Shell held a notorious secret meeting in Achnacarry Castle, "a romantic turreted mansion in Inverness-shire in the Highlands of Scotland". They agreed to run the world oil market as an international cartel. To avoid competition, existing market shares were essentially frozen and, for this reason, the agreement was known informally as "As Is".[xxxi]

These efforts were for a while complemented by US government measures to control the oil market and for many years the Texas Railroad Commission played a major part in regulating oil production in the US to help control US and international prices.[xxxii]

Whatever criticism may be made of the efforts of the, mainly Middle East, oil producers to manage oil prices through the OPEC cartel, it is hard to deny that they are part of a tradition reaching back to the earliest days of oil production.

[xxix] Frankel, P. *Essentials of Petroleum*. Frank Cass, 1946.
[xxx] Yergin, D. *Op. cit.*
[xxxi] Sampson, A. *The Seven Sisters*. Hodder and Stoughton, 1975.
[xxxii] Sampson *Op. cit.*

The precise consequences for the global economy of the spectacular oil price rises of the 1970s and early 1980s have been hotly contested. But there is little doubt that they made some contribution to the global recession that ensued. Ever since then, the price of oil has been recognised as a significant factor in the health of the global economy. Almost certainly the world economy is less sensitive to oil prices now than it was then. Nevertheless, especially when the global economy is fragile, the impact of oil price levels can certainly be significant.

Having reviewed the geopolitical and economic impact of oil in the past, we move on to consider the future and to discuss the factors that are profoundly shifting the world oil outlook today.

The first of these is technology. The oil and gas industries have made huge advances in recent decades. These include "3D Seismic" which makes it possible to picture underground rock formations in ever increasing detail before drilling is even started; horizontal drilling, which makes it possible to extract oil from wider areas from a single drilling site; and technologies for exploring in deeper waters and more extreme conditions. The ability to develop "oil sands" has opened up potentially vast reserves, mainly in Canada.

But it is the combination of horizontal drilling with "fracking" to release hydrocarbons from shale or other "tight" rocks that has had the greatest impact, mainly in the US and Canada. The technology is discussed more fully in the gas section below. Whereas conventional oil production only releases oil that has migrated underground from where it was originally formed and become trapped in porous rocks, fracking can release some of the much greater volumes of hydrocarbons that are still in their original "source" rocks.

Announcing "Project Independence" in November 1973, Richard Nixon declared, "From its beginning 200 years ago, throughout its history, America has made sacrifices of blood and also of treasure to achieve and maintain its independence. In the last third of this century, our independence will depend on maintaining and achieving self-sufficiency in energy".[xxxiii] For 40 years it was not to be. But now, largely as a result of

---

[xxxiii] Richard Nixon, Address to the Nation about Energy Policy, 25 November 1973. Available at http://www.presidency.ucsb.edu/ws/?pid=4051http://www.presidency.ucsb.edu/ws/?pid=4051. Accessed 7 July 2014.

shale oil and gas developments, the US, with Canada, is coming close to the holy grail of energy self-sufficiency. The impact on world oil markets has been huge. Between 2008 and 2013, US oil production has increased by more than the entire production of Iraq.

What impact, if any, will greater energy independence have on American foreign policy? This question has been much debated in Washington. Does it mean that the US will be less interested in the Middle East and that Pax Americana will no longer protect world energy trade? It is possible that there will not be much change. America will still be involved in world oil trade, exporting as well as importing, and will still take its oil prices, hugely important politically, from world markets. Energy and the Middle East will still have global strategic importance for the US, and of course Israel will still be an important ally. The Obama administration announced a "pivot to Asia" in its diplomatic and military policy, and this implied some degree of rebalancing.[xxxiv] The impact that Donald Trump's presidency may have on US policy in this area remains a wild card at the time of writing in June 2017.

The next tectonic movement has been the rise of China and other major developing nations not just as economic powers, but also as energy consumers and importers. This has shifted the focus of world oil trade from the Atlantic to Asia and the Pacific. As prosperity increases in China, the new middle classes want personal transport. Beijing has five ring roads and all of them are congested. China is now the world's biggest market for conventional internal combustion vehicles.

As a result, China has also now succeeded the US as the world's largest oil importer. One of the largest movements in world oil trade is now from the Middle East to China. China has become Saudi Arabia's best customer. This trend may be set to continue, because Asia has only 4% of world oil reserves. The tankers coming down the Straits of Hormuz now turn left instead of right.

It is a surprising fact that, although the Asia Pacific is now the largest oil trading region, the Brent field in Scotland remains the main benchmark for international oil prices. Brent is a declining field. The volumes of oil

---

[xxxiv] Jones, B. *et al. Fueling a New Order*, Project on International Order and Strategy, Brookings, March 2014.

are small and other fields in the UK sector of the North Sea have had to be included to keep the benchmark viable. Obviously the location of Brent, on the opposite side of the world, is not ideal for trading in Asia and concerns have been expressed, especially at times of high oil prices, that an "Asia premium" has arisen. A viable benchmark requires a good variety of arms-length buyers and sellers and a high degree of transparency. This is not always easy to achieve. However, an Asia benchmark may develop around oil trading in Singapore.

Vulnerability to oil supply interruptions is now a strategic concern for China. At present, it is the US that guarantees the security of the strategic trade routes, including the pinch points at the Straits of Hormuz and Malacca. China benefits from this. On the other hand, the US is a strategic rival. Dependence on Pax Americana may also be a matter of concern for China.[xxxv]

In her efforts to enhance oil security, China has invested in oil development and entered into government finance for oil agreements in many parts of the world. For the most part this oil is not shipped to China, but sold competitively on international markets. These investments make a positive contribution to world oil supply and, therefore, to the stability of the oil price. Without them it would be more difficult to cope with China's increasing demand. Nevertheless, as China continues to increase its acquisitions the risks of local resentment, and of tensions with Western interests, may increase. We have already seen some signs of this in relation to the international effort to bring peace in the Sudan and US oil sanctions against Iran.

China's claim to islands in the East and South China Seas that are also claimed by Japan, South Korea, and the Philippines is one of the most serious sources of international tension today. The likely presence of oil and gas reserves in the waters around these may or may not have motivated these claims in the first place but it certainly gives added weight to these disputes. It was the positioning of a Chinese oil rig in disputed waters that has most recently given rise to violent anti-Chinese riots in South Korea, as well as a sinking of a South Korean fishing vessel.

[xxxv] http://www.defense.gov/News/newsarticle.aspx?ID=119505.

We now turn to the pressures and uncertainties of the global oil market today. As I have described, there is nothing new about the volatility of oil prices and the uncertain balance between supply and demand. Political instability, and even conflict, in the Middle East remains, unfortunately, a continuing problem.

One often hears claims that "resource nationalism" is becoming an increasing problem in international oil markets, and indeed in energy markets generally. Governments have always sought to get the best possible value for their own people from their national resources. There is nothing wrong with that. However, in recent decades more of the oil producing nations have decided to develop their oil reserves through their own national oil companies (NOC). These NOCs now own most of the world's reserves and are also dominant in conventional oil production. This has been helped by the fact that much of the world's oil and gas technological expertise now resides with giant international oil service companies such as Schlumberger and Halliburton, and not just with the major oil companies themselves. It is difficult to generalise about these NOCs. Some are well managed, technically advanced, and relatively independent organisations. Others are the instruments of corrupt regimes.

For the stability of oil supply it is important that companies of all kinds should be able to invest in reasonably stable conditions. The seizures, in recent years, by the government of Argentina of oil assets belonging to the Spanish oil company Repsol and by Venezuela of assets belonging to EXXON and Conoco/Phillips have obviously had a negative impact on private investment in the region. The Energy Charter Organisation is a legally binding international agreement intended to provide greater security for international energy investments. The question of how to increase the influence of the Charter, and other international energy organisations, will be taken up in Chapter 10.

OPEC itself is a form of resource nationalism. Its impact over the years is very difficult to judge. When OPEC was first formed it brought to an end a quasi-colonial relationship between Middle East producers and the Western oil majors. Prices rose sharply, and OPEC countries have benefited from high oil prices over extended periods, including nearly five years of $100 plus prices that ended abruptly in 2014. It is hard to judge how large a contribution OPEC's own policies made to this but, to the extent that OPEC did contribute, it will have been largely due to Saudi Arabia. Saudi Arabia is the only OPEC member that has been willing, consistently, to produce less than its potential, thus maintaining a significant margin of spare capacity. The lower prices that have been experienced since 2014 can be seen as a rebound from OPEC's efforts to keep prices high for so long.

The West has developed a rather ambivalent relationship with OPEC in recent years. It is true that high oil prices are not in the interests of importing nations. But nor are very steep price fluctuations, as discussed in the following section. Saudi Arabia has shown itself willing, as a last resort, to draw on its spare capacity to respond to oil supply crisis or extreme price hikes. Probably, they believe that this is in the long-term interests of producers. It may also reflect Saudi Arabia's strategic links with the West. In practice, therefore, Saudi Arabia has become the first port of call in oil emergencies. The oil stocks held by the IEA are the main international reserve available for use in a crisis but the IEA would be unlikely to draw on them, today, without first consulting with OPEC and with Saudi Arabia in particular.

The elephant in the room of energy policy today is the 2014 collapse of oil prices, to below one third of their peak, and the partial recovery that has taken place in June 2017. Why did prices collapse so suddenly? The main factors are obvious. They have included the rapid growth of shale oil production in the US and of other kinds of non-conventional oil. Prices of $100 plus, well above the marginal cost of new production, inevitably led to an investment boom in all these areas. Another big factor has been the slowing of economic growth, and therefore of the growth of oil demand, especially in China. Also, vehicle efficiency has been improving around the world, driven by tougher government regulations. As a result, oil demand in the US and Europe is now static and, probably, starting a declining path. Also, oil production in Iraq has been recovering steadily.

Saudi Arabia's decision not to cut production further in order to accommodate these developments was the trigger of the price collapse, and there has been much speculation as to their motives. No doubt it is true that they have no great desire to prop up prices for the benefit of the US shale oil producers or to stimulate the recovery of oil production in Iran. But even Saudi Arabia, the world's largest producer, has only so much space to cut production without excessive loss of earnings, and in the end the Saudis seem to have decided that maintaining an acceptable market share took precedence over supporting the market price for the benefit of all producers. The Saudis may also realise that the global transition to low carbon energy, and the prospect of electric vehicles, mean that the future value of oil left in the ground is not quite as assured as they used to think. The effect of this Saudi decision was that OPEC, for the time being, was no longer a significant player.

Lower oil prices are generally regarded as being good for economic growth because they increase the spending power of middle income families. We have certainly seen this effect. But, contrary to what was hoped, they do not seem to have kick started a global economic recovery in the years after the 2014 price collapse. Perhaps, this was because there have also been negative effects. These include the cancellation of many major projects, and job losses, in the energy sector itself. They also include the blow to the financial sector as the credit of many energy businesses has been undermined. And the international investments of the oil producing nations, and their sovereign wealth funds, have been curtailed. All these negative shocks to the economic system have, perhaps, been greater than anticipated.

Oil prices at less than $40 per barrel were below the long-term marginal cost of production, which is probably in the range of $50–$70. No doubt oil prices will peak again, as the investment cycle turns. It is difficult to judge how soon. Will China enjoy a soft economic landing? When will the global economy recover from its current slow-down? What is the outlook for US shale oil production? When will Iran's oil production recover its former levels? What is the political and security outlook in the Middle East? As always, there are too many imponderables to make an oil price prediction. Even when the fundamentals have shifted, market psychology will play a part in determining the precise moment of price movement.

The latest situation, in June 2017, is that an agreement by OPEC and Russia to limit production levels has achieved a partial price recovery, to over $50 per barrel. This is an alliance of desperation. The three key players are Russia, Saudi Arabia, and Iran. Saudi Arabia and Iran are bitter rivals as leaders, respectively, of the Sunni and Shia Islamic sects.

Russia and Iran are on the opposite side from Saudi Arabia in the civil war in Syria. The agreement will need to be highly rewarding to all parties if it is to hang together. For the moment, that seems to be the case.

US shale production did not decline as sharply as Saudi Arabia no doubt hoped, when oil prices fell. The number of active drilling rigs declined by two thirds, but the remaining rigs were much more productive and US onshore oil production only declined by about 6% from its peak. There were technology gains and a lot of costs were taken out of the business. The big unknown is how fast production will rebound following the moderate increase in price. An optimistic view is that shale oil may have fundamentally improved the short-term price elasticity of oil production, which should mean somewhat more stable prices in the future. We shall see.

The volatility of oil prices is immensely damaging to the global economy, and especially to oil importing developing nations. A normal reaction might be to suggest that consumers and producers should get together to agree on a moderate and sustainable range for oil prices. This has been suggested many times. But it's hopelessly unrealistic because, even assuming that producers and consumers could agree on the range, which they can't, price variations are needed to correct imbalances to demand and supply; in other words to incentivise new investment and reduce demand when there is a shortage and vice versa when there is a surplus. Experience with cartels, including OPEC, has shown that trying to peg the price in defiance of the market will eventually lead to even greater instability.

Leaving aside the possible contribution of shale oil, the most promising way to moderate the volatility of energy markets is to improve the information available to investors, so that they can plan to bring on new production at the right level to match tomorrow's energy needs. The information needed includes the capacity and expected production profiles of existing fields, as well as investment plans and likely yield. Because most oil production and most oil reserves are now attributable to national governments and their state oil companies, this requires international cooperation at government level.

The best international effort to achieve this today is the Joint Organisations Data Initiative (JODI). JODI is a collaboration of International Energy Forum, OPEC, and the IEA, with other international statistical bodies, to gather and publish information on oil and gas production and investment. The G20 have given top-level support to this process. JODI is imperfect with regard to the timeliness, coverage, and quality of its statistics, but it's the best international initiative of its kind that we have.

More extreme than oil price volatility is the risk of an oil supply crisis. That may seem farfetched at a time when oil supply is in surplus. But the market will no doubt tighten again in due course. Because oil supply is so concentrated in the Middle East, and so much of world oil production passes through the Straits of Hormuz, the possibility of a serious disruption remains. The nightmare outcome would be a free for all amongst major importers with prices spiralling even above levels of recent years and serious hardship for the weakest players. In Chapter 3, I will discuss international efforts to manage oil supply crises in the interests of all.

The question of "peak oil" was a major topic of energy debate in the 1970s and 1980s. Production of oil from conventional oil fields has probably already peaked. To that extent the "peakers" can claim victory, but it is a hollow triumph. Conventional oil fields, where oil has migrated underground from where it was formed and been trapped in porous rocks under non-porous cap rocks, represent only a small part of the oil in the ground, albeit the most easily produced. Associated oil production from gas fields and unconventional oil, including shale oil and oil sands, has been growing rapidly and total oil production continues to increase. For practical purposes the amount of oil in the ground can be regarded as limitless. Oil availability depends on how much we are willing to pay to extract it and the march of oil development technology, which has been impressive in recent decades. Peak oil will be determined by demand as much as supply. As the saying goes, "the stone age didn't end because they ran out of stones".

Today the boot is on the other foot. The concern today is not so much that the oil might run out as that the oil we have already discovered is much more than can be safely used without damage to the climate. Analysts have pointed out that the carbon in the published reserves of fossil energy companies, including oil companies, far exceeds the limit for global emissions if we are to limit global warming to two degrees.

Climate change policies are starting to have an effect on oil markets. As already mentioned, the drive to enhance vehicle efficiency has definitely contributed to the flattening of oil demand in the US and Europe. International efforts, led by the G20, to reduce fossil fuel subsidies may also moderate demand growth, especially in Asia and the Middle East. The uptake of electric or hydrogen cars has been modest and the oil industry does not see them as an immediate threat to their business. The industry may be wrong about that. This topic, which is fundamental to the medium-term outlook of the oil industry is considered in depth in Chapter 6.

It is important to realise that passenger cars only make up about a quarter of oil world's oil demand.[xxxvi] Another quarter is made up of trucks, aircraft, and ships, for all of which alternatives are more difficult to find. Nearly 20% is used in industry, where substitution may also be difficult. Just over 10% is used for petrochemical feedstocks, which are not necessarily a source of carbon emissions.

Oil remains, for the time being, central to the global energy economy and a considerable factor in geopolitics. International cooperation to protect energy security and to promote open, transparent and, as far as possible, stable, energy markets remains vital. We need to adapt the institutions for this cooperation to better reflect the rise of China and other developing nations. Inherent in the current low level of oil prices, and the cancellation of major investment projects is the risk of a price rebound in the future. That may be inevitable. But we must try to make good quality information available to investors, and to producing and consuming countries, about the outlook for the demand and supply balance. This includes the impact that the low carbon transition may have on oil demand in the medium and longer term. We also need, as far as possible, to provide stable opportunities for international energy investment. The need to reform international energy institutions to make them more inclusive and to provide a better platform for achieving these objectives is a topic of Chapter 10.

5. Natural Gas

Natural gas used to be a junior partner to oil and coal. The discovery of gas reserves was a nuisance to companies who were primarily looking for much more lucrative oil. But all that has changed. Gas is projected to be

---

[xxxvi] International Energy Agency, *World Energy Outlook*, 2016. *Op. cit.*

the fastest growing of the fossil fuels over the next 20 years, and over that period its contribution may almost catch up with those of the other two. The IEA has been speculating about the coming of a "golden age of gas".[xxxvii] Now that the industry is actively looking for it, gas has turned out to be surprisingly plentiful and widespread. The development of trade in gas in liquefied form, discussed below, means that gas that used to be regarded as stranded can now be put on the world market. The combined cycle gas turbine for electric power, discussed further in Chapter 6, and improvements in domestic gas boilers have made gas a highly efficient fuel. Its costs are competitive and it may have considerable environmental advantages over other fossil fuels.

Gas is a sort of environmental half-way house. Gas combustion to generate power produces only about half the carbon dioxide emissions of coal. However natural gas is, itself, a much more virulent greenhouse gas than carbon dioxide, though it does not stay in the atmosphere for so long. The climate change benefits of switching from coal to gas depends absolutely on the careful management of gas production and transmission to avoid leaks to the atmosphere. It is a matter of concern that President Trump appears willing to relax US regulation in this area. Gas is also much less polluting of the local environment than coal. The role that gas can play as a "transitional" fuel, and perhaps even for the longer-term, in reducing greenhouse gas emissions will be discussed in Chapter 6.

Gas can also help to reduce local pollution of the atmosphere. London's famous yellow smog was romanticised in the Sherlock Holmes stories and elsewhere:

> ... a dense drizzly fog lay low upon the great city. Mud-coloured clouds drooped sadly over the muddy streets. Down the Strand the lamps were but misty splotches of diffused light which threw a feeble circular glimmer upon the slimy pavement. The yellow glare from the shop-windows streamed out into the steamy, vaporous air, and threw a murky, shifting radiance across the crowded thoroughfare.[xxxviii]

[xxxvii] International Energy Agency, *Are We Entering a Golden Age of Gas?* World Energy Outlook 2011 — Special Report.
[xxxviii] Sir Arthur Conan Doyle. *The Sign of Four*, 1890.

But this pollution from the coal fires that heated all London homes at the time was, in reality, a serious health hazard. It has been estimated that the severe 1952 London smog caused some 12,000 deaths.[xxxix] London's yellow smog, though not, of course, all its air pollution, was eliminated by the widespread adoption of gas for domestic heating and local industry in the 1960s and 1970s. Switching from coal to gas for industry and home heating now offers one of the most promising options for cleaning up the air pollution in cities such as Beijing and Delhi.

The countries with the largest reserves of gas are Russia, Iran, and Qatar. There are substantial reserves in the Middle East, in the Central Asia, especially Turkmenistan, in Africa, in Australia, and in the US. Western Europe is not so well endowed, with the exception of Norway and, to some extent, the UK.

Gas has traditionally been transported through pipes and this has meant that its markets have been national or regional rather than global. It has also meant that the history of gas markets has largely been about pipeline development and regulation. In recent decades, however, the shipping of gas in liquid form (LNG) has become much more common and it is this that is opening up a global market in gas that is much more analogous to oil. Cheap shale gas production in the US, which has already revolutionised the US gas market, is set to have a growing impact on world markets.

The US has developed, by far, the world's most competitive natural gas industry. But that was not the original intention. In 1938, when regulation started, it was considered that the ownership of major gas pipelines conferred a "natural monopoly". The government controlled the prices that the pipelines could charge for gas transmission and also restricted the construction of new pipelines and eventually regulated the price of gas itself.

However, by the late 1970s, this was leading to gas shortages. Gas price controls were lifted, and the pipeline companies responded by entering into long-term fixed volume (take or pay) purchase contracts with gas

---

[xxxix] Bell M.L. *et al. Reassessment of the Lethal London Fog of 1952*, Environmental Health Perspectives, June 2001; 109(Suppl 3): 389–394.

producers. The new pricing freedom quickly led to an oversupply, known as the gas "bubble" and then the gas "sausage" as it persisted.

When I arrived in Washington in 1985, as energy counsellor at the British Embassy, the gas industry was in turmoil. Gas was available at prices well below those in the take or pay contracts but the customers, who included big business as well as municipal distribution companies, could not buy it because the pipeline companies controlled access. The regulator, the Federal Energy Regulatory Commission (FERC), issued order 436, which gave the pipelines the option of stepping out of the gas market and simply selling transmission. But they were unwilling to do so because that would have enabled their customers to undercut the long-term contracts to which they were committed. There was a lot of frustration.

Finally, in order 636, the FERC issued a judgement of Solomon. The pipelines were required to "unbundle". That is to say, they were required to separate their transmission businesses from their interests in buying and selling gas. Pipeline services were to be open to buyers and sellers of gas on an equal basis. This left the take or pay contracts, which now appeared overpriced, stranded. Many of the pipeline companies would have been bankrupt. Therefore the 636 order gave them the right to load some of the costs of unwinding these contracts onto their transmission charges.[xl]

This highly successful US regulatory development is important because it offers a model for the "deregulation" of energy industries that have been taking place all over the world. The aim is to open up new market opportunities and increase competition. Some of the success of FERC 636 has been due to the depth and diversity of the US oil and gas drilling industry and its highly developed pipeline network. The US now has the world's most liquid and transparent gas market, located at the Henry Hub in Louisiana, a centre for gas transmission that connects with nine interstate and four intrastate pipelines.

[xl] Natural Gas.org, website of the US Natural Gas Supply Organisation, *The History of Regulation; Natural Gas.* Available at http://naturalgas.org/regulation/history/. Accessed 9 July 2014.

FERC 636, together with research originally conducted in US national laboratories, can now be seen to have opened the door to the subsequent spectacular development of the gas industry in the US.

The most important recent development in gas technology has been production of gas, as well as oil, oil from relatively tight shale rocks through horizontal drilling and hydraulic fracturing, or "fracking".[xli] Fracking is a process in which oil or gas is released by pumping fluid under very high pressure to create cracks in the bearing rock. The fluid is mainly water and sand but contains other additives designed to ensure that the sand penetrates the cracks and keeps them open.

Fracking has become controversial mainly because of concerns that it may lead to the pollution of aquifers or the release of methane, which is a powerful greenhouse gas, to the atmosphere. In most situations, the location of the fracking, more than a kilometre below ground, makes it very unlikely that it can directly pollute aquifers. The main risk to aquifers arises from the integrity of the vertical section of the well, a risk that is not different in principle from that arising from a conventional oil or gas well, except for the fact that fracking requires many more wells to produce the same volume of oil or gas. There is no doubt that serious methane leakage to the atmosphere has occurred from fracking operations in the US. But well managed fracking operations can avoid this, at some additional cost.

Fracking is an industrial process that involves heavy equipment and hazardous materials. It is not suitable for all locations. But provided that it is well managed and regulated, it need not impose exceptional environmental risk. To a large degree this is a question of trust in managers and regulators.

The other concern about fracking is that it opens up the potential for a large new source of fossil fuels at a time when we need to reduce carbon emissions. The key question is, what will the additional gas displace? If it displaces other higher cost gas, the consequence is essentially neutral for climate change. If it displaces coal, as it has largely done in the US and would probably do in China, the consequence is positive. If it displaces renewables or leads to additional energy consumption then it will cause

---

[xli] Hirst, N. *et al. Shale Gas and Climate Change*, Grantham Institute for Climate Change, Imperial College London, Briefing Paper No. 10, 2013.

additional emissions. A study that the IEA carried out in 2012 suggested that an increase in gas supplies would lead to modest reductions in total carbon dioxide emissions 2035. But there are many uncertainties in this calculation.

The shale gas revolution has contributed to spot US gas prices that, in 2015, were less than half those in Europe and about a quarter of those in Asia, with consequent benefits for living costs and industrial competitiveness, though this difference had reduced sharply by 2016. It has reversed the previous expectation that the US was to become a substantial gas importer and contributed to the trend towards greater energy independence for the US. President Obama claimed that hydraulic fracturing could create more than 600,000 jobs in the US. One consequence has been a decline in coal-fired electricity generation in the US. This contributed to reducing US carbon emissions in the first quarter of 2012 to their lowest level since 1992. However it has also contributed to an increase in US coal exports.

So far shale gas production has largely been confined to North America, but recoverable shale gas resources are widely spread across the world. According to initial assessments by the US Energy Information Administration technically recoverable shale gas resources could be sufficient to increase conventional gas resources by nearly 50%. China, Argentina, Algeria, the US, Canada, Mexico, and Australia may be the best endowed. As with shale oil, it is difficult to judge to what extent other nations will be able to follow the American example on shale gas. The conditions are not the same elsewhere, but substantial potential undoubtedly exists.[xlii]

With faltering steps, the European Union has attempted to follow in the footsteps of the US to liberalise its gas market. Since privatisation of the state-owned gas industry in the 1980s, the UK has developed a gas trading hub, the "National Balancing Point" and a liquid market in gas. However, progress in continental Europe has been much slower. It has proved difficult to unravel the positions of powerful national quasi-monopolies who have bought gas from Russia on long-term "take or pay" contracts. These contracts are not competitively priced against other gas sources but are indexed to the price of oil. The argument is made that these

---

[xlii] Hirst, N. *et al. Ibid.*

long-term contracts were essential to enable Russia to finance the development of new fields and pipelines. This has some merit. There is a difficult dilemma that all gas regulators face that some degree of temporary monopoly of the use of new gas facilities may be necessary to bring about investments that will be of universal benefit in the future. As of 2017 a well-supplied international market in LNG is putting increasing pressure on the Russian gas supplier, Gazprom, to liberalise contract terms. Hopefully, Europe will continue its progress towards a genuinely competitive gas market.

Europe's conventional gas reserves, mainly located in Norway and the Netherlands, are less than half those of the US, but Europe also has considerable shale gas potential.[xliii] Many believe that more competitive European gas markets could open the door to more diversified and competitive supply. In 2017, Europe's gas demand is declining due to increased use of renewables and even coal for power generation.

Russia is believed to have the world's largest gas resource, about a quarter of the world's proven reserves.[xliv] The Russian gas sector is still largely state run through Gazprom, which controls most of Russia's gas reserves and gas production and also has a monopoly of gas exports, though that may be ending soon.[xlv] In Russia, gas sold at well below international prices, is by far the largest domestic source of energy. So far, Russia has mainly relied on gas reserves in Western Siberia and its main export market has been in Europe. Now, Russia hopes to open up export markets in China and Japan.

Exports from Russia account for 30% of Europe's gas supply and Europe is the destination of three quarters of Russia's gas exports.[xlvi] This degree of dependency is unwelcome to both parties but, for the time being, Russia and Europe are locked in their gas trade embrace. The relationship has survived even the recent crisis in Ukraine. Europe has imposed a variety of sanctions on Russia, but ending the gas purchases is not an option.

[xliii] US EIA, *Statistics, Op. cit.*
[xliv] US EIA, *Russia Country Report*. Revised March 2014.
[xlv] US EIA, *Russia Analysis.*
[xlvi] International Energy Agency, *Medium Term Gas Market Report*, 2014.

In the longer-term, both parties are looking for means of diversification. Russia is opening up its gas trade to the East, mainly to China. The recent signing, after more than a decade of fruitless negotiations, of a 30 years agreement to supply gas worth an estimated $400 billion to China should contribute to that. So far, Russian gas production has largely been in Western Siberia. However, the opening up of markets to the East will require the development of difficult and relatively high cost reserves in the Yamal Peninsula, in the Arctic, in the East of Siberia, and on Sakhalin Island, off Russia's east coast. Development of these reserves, which is one of the greatest imponderables in world gas supply, will have to compete with international LNG supplies, including possible low-cost exports from the US, and, possibly, shale gas production in China itself.

Europe is looking at possible pipelines to bring gas from the Caspian region, and tanker carried LNG from the Middle East and Africa to diversify its gas supplies. But this will take time.

In China, gas is still the poor relation to coal and oil, contributing only 5% of energy supply in 2014. But this is may be about to change, as China comes to grips with the problems of urban energy pollution. China is ranked 11th in the world for conventional gas reserves, but these are fairly modest in relation to China's huge population. Already nearly a third of China's gas is imported, and about half of the rapid increase in gas demand expected in the next few years will probably come from imports.[xlvii]

China is believed to have the world's largest shale gas reserves, although the geology may not be as favourable as in the US. It is still very uncertain whether China could achieve a shale gas revolution remotely similar to that in the US and this could be one of the biggest wild cards in the world energy outlook. China has significant potential to release gas from underground coal beds and there are also believed to be large gas reserves under the East China Sea and the South China Sea, some of them in disputed waters. China is reforming its state controlled gas market to introduce more competitive pricing. The future of China's domestic gas industry will depend on the continuation of these reforms to create a more efficient market and a more diversified industry.

[xlvii] International Energy Agency, *Gas Market Report, Op. cit.*

One of the most significant developments for world gas in recent years has been the discovery of huge new reserves in the Caspian region and central Asia, including the vast South Yolotan gas field, the second largest in the world, in Turkmenistan in 2011.[xlviii] Both China and the EU were interested in this gas. But many have commented on the contrast in the effectiveness of their responses. China has signed a major gas supply deal with Turkmenistan and has already, in 2017, built a thousand mile pipeline to take delivery of the gas. Europe is still struggling with the various national and commercial interests involved in taking the gas from the Caspian region to the West.

The technology for transporting natural gas in LNG in tankers has been around for a long time. But it is only in the last couple of decades that it has started to play a really large international role. Now, it is transforming the world gas industry by liberating major gas reserves with no pipeline outlet. LNG export capacity already exceeds 10% of world gas demand and is set to increase by more than 25% over the next five years. Qatar has been the largest exporter but is now set to be overtaken by Australia, where seven huge developments are about to come on stream. Africa, Indonesia, and now the US, all have major export potential. Japan, China, and Europe are amongst the largest importers.

Since the 2011 earthquake and Fukushima nuclear accident, Japan has had to close all its nuclear power stations, which were providing 30% of its electricity. Most of this has been replaced by urgently contracted gas in the form of LNG, making Japan by far the world's largest LNG importer today. Fortuitously, this has coincided with the rapid build-up of shale gas production in the US. LNG that was originally intended to supply the US has been diverted to Japan. Indeed, the terminuses that were built to handle US imports are now being reconfigured for the export trade.

In theory, the ability of LNG to be traded across the globe should lead to a global market in gas on the same lines as oil. Gas prices in Asia, and to a lesser extent Europe, have been higher than those in the US to an extent that is not fully explained by the costs of LNG shipment and

---

[xlviii]The Telegraph, *Second Largest Gas Field Found in Turkmenistan*, 11 July 2014. Available at http://www.telegraph.co.uk/finance/newsbysector/energy/oilandgas/8535672/Second-largest-gas-field-found-in-Turkmenistan.html. Accessed 11 July 2014.

processing. The situation eased as oil prices fell after 2014, since most Asian imports have been purchased on long-term contracts with prices indexed to oil. Also, Japan has begun to restart some of its nuclear power stations, the new Australian capacity has started to come on stream, and exports from the US have become a reality. By 2016, the differential of gas prices between Asia and the US was only a fraction of what it had been in 2013. Perhaps the global gas market is beginning to work.

Nevertheless, international LNG markets are far from perfect and this remains a concern, especially for China. As with oil markets, there is a need for a trading hub in Asia. China's sense of being an outsider in world energy trade, perhaps less pressing now that oil and gas prices have fallen, is nevertheless a significant topic for world energy cooperation, something that is taken up again in Chapter 10.

6. Conclusions

This chapter has been about fossil fuels and the benefits that they confer. We now know that the use of these fuels is causing irreparable and potentially disastrous damage to the atmosphere and that we must drastically reduce our reliance on them, unless we are willing to capture and store their emissions. A large part of the rest of this book is concerned with the question of how we can achieve this reduction without losing those benefits that we, in the developed world, mostly enjoy and that many others aspire to enjoy.

For several decades to come, at least, fossil fuels will play a major and essential part in meeting our energy needs. Global prosperity and peaceful development will continue to depend on the efficient functioning of fossil energy markets. That means that international cooperation to keep energy markets open, to improve their transparency and stability, and to manage crises, will still be important.

Our continuing dependence on fossil fuels during a transition period also means that our efforts to address climate change cannot just be focused on low carbon alternatives but must also include the right incentives for fossil fuels themselves to be used in the cleanest and most efficient way. There needs to be a better understanding between the fossil fuel industry and governments about the transition process. The narrowing of their markets will never be welcome to the industry. But what the industry have called for again and again is a degree of clarity about government policies that can enable them to plan in a realistic way for the future. There are various ways of achieving this. But a clear and consistent policy to adopt steadily rising carbon pricing, reflecting the damage that emissions do to the environment, would provide the best basis for enabling the fossil industries to plan for the future. Some of the leading companies are already calling for this.

There are difficult questions of equity, or "energy justice" as some prefer, in the process of running down fossil fuel dependence. Can we expect developing nations to rule out fossil energy even if, in some cases, it would be part of the cheapest and most accessible development route? Is the developed world willing to pay the additional cost of low carbon pathways? Perhaps continuing falls in the costs of renewables and energy storage will eliminate this as a problem. But we haven't reached that stage quite yet.

Nobody can predict the longer-term future. Renewables are achieving spectacular cost reductions. They are plainly set to play a major and possibly dominant role in world energy, and especially in electricity generation. Governments are right to promote them. But there are other areas, including heat, industrial processing, and transport, where the pathway is not yet so clear.

We should not assume that fossil fuel costs will always increase. There will probably continue to be locations where coal is relatively cheap. The shale gas revolution has demonstrated that it is not only the renewables industries that are capable of cost reducing technological progress. Oil may cost around $50 per barrel in 2017, but a large share of the world's oil reserves are capable of production at costs below $10 per barrel. We should not rule out the possibility that a substantial fossil fuel industry, capturing and storing its emissions, may be an essential part of the global low carbon economy of the future. Government and industry need to work together to make sure that that option remains open.

Chapter 3

Energy Security and Emergency Planning

1. Introduction

Energy security is one of the three main objectives of energy policy, together with affordability and protection of the environment. When supply is assured it may disappear from day-to-day concerns, but as soon as there is a credible threat to energy supply it becomes an absolutely overriding concern of governments. Emergency planning exercises are usually rather obscure things, run by not very senior staff. But if the balloon really goes up, you can be sure of the attention of the prime minister and his or her closest aides, because their political futures, if any, are on the line!

There is one very practical reason for this. People in developed economies regard the maintenance of secure energy supplies as one of the first duties of government. Experience shows that, unless there has been overriding *force majeur* in the form of a natural disaster, political leaders pay a heavy price for supply failures.[i]

---

[i]Royal Academy of Engineering, *Counting the Cost: The Economic and Social Costs of Electricity Shortfall in the UK*, 2014.

2. Electric Power Failure

Let me give some examples of electric power disruptions. In 1996, the state of California passed legislation to deregulate its electricity industry. Control of the transmissions system passed from the three major utilities to an Independent System Operator. These utilities were also encouraged to sell off some of their power stations to promote competition. By July 2000, an acute power shortage had developed. The exact causes of this have been hotly disputed. However, rising gas prices, reduced hydroelectric power as a result of drought, flaws in the complex structure of the deregulation including price controls, and market manipulation by the electricity suppliers are all considered to have played a part.

During 2000 and 2001, there were a series of "rolling blackouts" affecting, at their worst, almost a million people at a time. The crisis was estimated to have burdened California consumers and businesses with some $40 billion of additional energy costs, as the state became a distress purchaser of power. If one adds the costs of the blackouts and reductions in economic growth, "conservatively, the total costs can be placed around $40 billion to $45 billion or around 3.5% of the yearly total economic output of California".[ii] It was a big hit. The state of California was put on "credit watch" by the rating agencies. Governor Davis was "recalled"; that is to say, he was voted out of office before the end of his term, becoming one of the first state governors to suffer this fate. There is a heavy political and economic price to be paid for the failure to maintain electricity supply security!

In 1974, mainly as a result of the union-imposed working restrictions in the coal mining industry, the UK government led by Edward Heath imposed a "three day week" on commercial users of electricity in order to conserve coal supplies.[iii] Because these arrangements were carefully planned and announced, there was no social unrest and the impact on the economy was considered at the time to have been surprisingly modest.

[ii] Weare, C. *The California Electricity Crisis: Causes and Policy Options*, Public Policy Institute of California, 2003.

[iii] Worthington, D. *Looking Back at the Three Day Week.* 2014. New Historian. Available at http://www.newhistorian.com/looking-back-three-day-week/2405/. Accessed 23 November 2015.

Seeking to pin the blame on the unions, Heath called a general election with the slogan, "Who Governs Britain?". But he did not convince the electorate, who voted him and his Conservative party out. This episode had a profound effect on the future course of energy policy. In 1984 Mrs Thatcher, another Conservative prime minister, succeeded in facing down a national strike by the National Union of Mineworkers (NUM) partly as a result of a careful strategy of building coal stocks, without resorting to cuts in electricity supply, and went on to win a second general election in 1987.

In 2011, following the tsunami and nuclear power disaster at Fukushima, Japan faced the sudden loss of 30% of its generating capacity that derived from nuclear power. The government launched an appeal for power saving, called *setsuden*, which seems to have been extraordinarily successful. Offices and factories turned up thermostats and turned off lights. They cut back on overtime and shifted working hours. Consumers limited the use of air conditioners and turned down the brightness of TV screens. Dress codes were relaxed and people took to wearing lighter clothing at home and in their offices so that less air conditioning was needed. One survey reported that 86% of those who responded were taking energy-saving measures at home. The result was a reduction in electricity demand of about 15%, making up half the power that had been lost. This was sustained for several years and there are even hopes that *setsuden* may have a longer term impact on Japan's energy culture. All this was a remarkable achievement, considering that Japan was already one of the most energy-efficient countries in the world.

The success of *setsuden* no doubt reflected a widespread recognition that the nation faced a severe crisis of energy supply. Japan is a highly cohesive society with a strong sense of national unity, and perhaps this campaign would not have been so successful in a more heterogeneous culture. Nevertheless, it shows what can be achieved where there is full public support for crisis management measures.

When cuts in electricity supply become unavoidable, they must be well planned, shared across the population in a fair way through "rolling blackouts", and well communicated in advance. Where these conditions have been met, the process can be orderly and the economic and social consequences need not be catastrophic. But the political consequences, and the loss of confidence in the government of the day, are still likely to be severe.

The traditional threats to power security have arisen mainly from inadequate generation supply, equipment failure, and unexpected outages. But cyber security is now becoming an increasingly important issue. According to Christoph Frei, the Secretary General of the World Energy Council,

> Cyber threats are among top issues keeping energy leaders awake at night in Europe and North America. Over the past three years, we have seen a rapid change from zero awareness to headline presence. As a result, more than 30 countries have put in place ambitious cyber plans and strategies, considering cyber threats as a persistent risk to their economy.
>
> What makes cyber threats so dangerous is that they can go unnoticed until the real damage is clear, from stolen data over power outages to destruction of physical assets and great financial loss. Over the coming years we expect cyber risks to increase further and change the way we think about integrated infrastructure and supply chain management.[iv]

Businesses considering new investment are likely to regard secure electricity supply as one of their criteria in order to avoid the additional cost of back-up generation. More generally, the reliance of modern economies on the continuity of electricity supply may be increasing as information technology plays an ever greater part in everyday life, including the means of exchange and communication, and all kinds of operating systems. As the growing reliance on variable sources of electricity supply requires highly complex balancing systems, the stakes, measured in terms of the consequences of an unplanned power failure, are also rising.

3. Oil Security

It is the need to maintain the security of oil supply that has had the biggest impact on national and international politics.

[iv] World Energy Council, *The Road to Resilience: Managing Cyber Risks*, September 2016.

Oil supply crises take different forms. Some would regard it as a crisis when, as in 2008, an imbalance emerges between demand and supply on world markets causing a sharp increase in price. However, we are concerned here with the possibility of actual supply shortages. The most obvious causes of these, historically, have been deliberate embargoes or interruptions due to international conflict, civil unrest, or civil war. In Chapter 2, I described the crisis in oil supply that arose in the West as a result of Arab embargoes at the time of the Arab/Israeli war of 1973, sometimes regarded as the worst crisis to afflict the Western alliance since its foundation after World War 2.

In a crisis of this kind, the risk that oil-consuming nations face from the behaviour of other consuming nations is just as great as the risk arising directly from the embargo. Nations facing serious concerns about their oil supplies are apt to become desperate and competitive, and a free for all in which each nation tries to corner the limited supplies that are available is the worst possible outcome. Other consumers can become a worse threat than the producers who are responsible for the original shortage.

It was partly for this reason that Henry Kissinger, the US Secretary of State at the time, took the initiative to found the International Energy Agency (IEA) in 1974 as a club of major consuming nations committed to cooperation in the face of the Arab embargo. The IEA has since developed in many ways but at its heart, and still enshrined in its legally binding treaty, is a conceptually simple agreement. Members committed to maintain emergency oil stocks equivalent to 90 days of their imports, and, when an emergency is declared, to make coordinated drawings from these stocks and reductions in oil consumption calculated, collectively, to meet the shortfall. They also agreed to share the oil available on international markets equitably.

Since its foundation, as described in Chapter 10, the IEA has evolved to become the world's leading intergovernmental body addressing a full range of energy topics. Nevertheless, it retains its crucial role in managing oil emergencies. IEA member countries, essentially all the developed, Organisation for Economic Co-operation and Development (OECD) nations, now hold about 3.9 billion barrels of oil in their strategic stocks.[v]

---

[v] IEA, *Fact Sheet on IEA Stocks and Emergency Response*, 2004.

Even at $50 per barrel, this stockpile, which includes government held stocks as well as industry stocks held to meet government obligations, is worth about $200 billion, and is an immense investment in energy security by any standards.

With the single exception of Hurricane Katrina, which significantly disrupted US oil supplies in 2011, all major world oil supply disruptions since the 1950s have been associated with wars and instability in the Middle East. The greatest losses arose from the Arab–Israeli war and the Arab oil embargo of 1973–1974, the Iranian revolution of 1978–1979, the Iran–Iraq war of 1980–1981, and the Iraqi invasion of Kuwait in 1990. Each of these resulted in a loss of supply, at the peak, of more than 4 million barrels per day.[vi]

The IEA Treaty contains a formal legal mechanism for activating emergency response. This mechanism is cumbersome and also unrealistic, in that it requires governments to manage commercial crude shipments. Consequently it has never been used. However, the IEA subsequently developed a much more flexible mechanism which makes very quick action possible where needed. Since its inception the IEA has acted on three occasions to bring additional oil to the market through coordinated action; in response to the 1991 Gulf War, to Hurricane Katrina in 2005, and to the disruption of Libyan supplies in 2011.

There is a growing debate around the suitability of the IEA's stocking rules for today's needs. This broke the surface in 2012 when the US called for an IEA stock draw to moderate oil prices in the run-up to their presidential election but the IEA refused.[vii] The IEA's firm rule has always been that the stocks are there to address specific shortage in the oil market and not to attempt to manage oil prices. However, from the point of view of the US, as it becomes increasingly self-sufficient in oil, price and not supply is the issue. Indeed the US government recently agreed, as part of a budget deal, to sell down 58 million barrels of oil from its 695 million

---

[vi] International Energy Agency (IEA), *Response System for Oil Supply Emergencies*, 2012.
[vii] Reuters Business News, 16 March 2012, *US and UK Set to Agree on Emergency Oil Stocks Release*. Available at http://www.reuters.com/article/2012/03/16/us-oil-reserves-idUSBRE82E0UM20120316#zAud5U1rqOef26Vc.97. Accessed 27 November 2015.

barrels of reserves from 2018 to 2025.[viii] Admittedly these reserves are in excess of the US's obligation to the IEA. Nevertheless, this action highlights the debate that is going on in the US about what the Strategic Petroleum Reserve is really needed for in today's circumstances.

It is not just in the US that the IEA's reserve strategy is being questioned. The issues that arise are whether the 90-day stocking requirement remains appropriate for all members and how the IEA can better coordinate with non-members in the Asia-Pacific region, especially China and India.

One IEA member, Australia, is not in compliance with its stockholding obligation although it is understood to have plans to comply in the future. Australia's own strategy for handling an oil emergency is the first to allow higher oil prices to reduce demand and then to apply increasingly onerous regulatory restraint on oil demand including, if eventually required, retail petrol rationing.

The pattern of world energy trade has changed completely since the 1970s, as described in Chapter 2. Today 75% of oil exports from the Middle East are destined for Asia, whereas Europe is increasingly supplied from Russia and Africa, and the US from its own production as well as from Canada and Latin America. The greatest risk to energy security is in the Asia/Pacific region, and especially to China and India, who between them account for more than half the oil imports in the region. So it is unfortunate that the membership of the IEA, does not include these major emerging nations. The great majority of the IEA's stocks are held in Europe and North America, whereas the main problems are likely to arise on the other side of the world.

Chatham House conducted a study of Asia's oil security in 2014. They noted that 50% of Asia's oil supply came from the Middle East, in contrast to the US (12%) and Europe (16%):

Although Asia is more dependent on oil supplies from the Middle East than are the US or Europe, it is less organised to deal with disruptions of those supplies.... For some Asian countries even a relatively short

---

[viii] Bloomberg Business, *US Plans to Sell Down Strategic Oil Reserves to Raise Cash*, 27 October 2015, Available at http://www.bloomberg.com/news/articles/2015-10-27/u-s-plans-to-sell-down-strategic-oil-reserve-to-raise-cash.

supply disruption, accompanied by a large price increase, would pose major economic challenge … Governments are bound to intervene in the event of disruption … conflicts between government interventions would be likely — unless, that is, governments were to take account of the interests of other countries in the region.[ix]

The need for better coordination is obvious. Most realistically, this should be built around the IEA's emergency response mechanisms to which Japan and South Korea already belong. At their meeting in November 2015, IEA Ministers asked the Secretariat to "examine potential options for broadening the collective oil security mechanism" and to report back. So it appears that the IEA is now actively considering at least some of these issues.

4. Managing an Oil Crisis

The most acute threat to the UK's oil supplies in recent years came not from international events but from lorry drivers and farmers protesting against the level of petrol taxes. In 2000, they successfully blockaded four essential distribution depots, causing an immediate shortage.

The government team managing this crisis was suddenly confronted with the reality that almost all aspects of modern society depend on a continuing supply of oil. Oil is needed to get staff, such as sewage plant operators, doctors, nurses, gas engineers, power plant operators, prison officers, to work. It is needed to deliver groceries and medicines, and to supply factories and food processing plant. The list feels endless in such a situation, and the consequences of an actual failure of oil supply, even for a fairly short period, can be catastrophic.

An oil crisis is like a bank run. If the public senses that supplies will not be assured they will fill up their tanks wherever they can, and this can empty the reserves held at petrol stations in a few hours. To control this, some countries have petrol rationing systems, including pre-printed coupons, or electronic equivalents, ready for implementation at very short notice. Britain didn't have that. Such systems are expensive to maintain and keep up to date.

[ix] Chatham House, *Asia's Oil Supplies: Risks and Pragmatic Remedies*, May 2014.

Various suggestions were made. A maximum could be set on the amount of petrol that motorists could purchase at one time, but that might just lead to multiple visits to petrol stations. Another suggestion, which seems superficially perverse, was to set a minimum amount — forcing motorists not to fill up until their tanks were reasonably low.

The system that was ultimately proposed was to give priority to key workers, to be validated by letters from senior people in their organisations. Most petrol stations would be required only to serve people with these letters. It could have been done very quickly, though it remains unclear how well it would have worked in practice.

Many of the public rather sympathised with the lorry drivers and farmers over high petrol taxes, so the government would not have enjoyed much public sympathy.

Fortunately, the drivers who were blockading the depots came, at last, to understand the grave consequences of what they were doing, partly as a result of discussions with doctors whose hospitals were on the point of failure. The blockade was called off before the plans could be put into effect. The government did not directly back down, but there was a pause in the steady increases in petrol taxes that had been taking place. About 70% of the price of petrol in the UK is tax, and it seems the lesson for the government in this episode was an understanding of what the acceptable upper limit was for the public.

Our crisis was unusual because it cut off access even to the UK's own reserves, creating an acute and immediate shortage. In crises arising from international supply problems, these reserves would be available as a cushion. So there would be a bit more time. It is important to have well-prepared plans for dealing with oil supply crises of whatever kind. The most important element of these plans is the communication between the government, the oil companies, the police, and relevant sources of international information, especially the IEA. All these organisations need to have a credible and consistent line for the media at the very earliest stages of the crisis. Otherwise there will be panic. Any action to manage petrol supply has to be coordinated between the government, the oil companies, and the police. In the case of the blockading farmers and lorry drivers, the oil companies made it clear that their station staff could not be expected to manage any rationing system without a police presence at the petrol stations.

5. Saving Oil in a Hurry

Petrol rationing is extremely unpopular and, in some liberal societies, almost unworkable. So what other options are there for reducing demand in the event of a crisis? The IEA published an excellent study of this back in 2004.[x]

This study was prepared at the time of the oil price spike of 2004. The senior national representatives met at the IEA in Paris and wrung their hands over the severe consequences of the price spike for their economies. Why couldn't the Organization of the Petroleum Exporting Countries (OPEC) nations be persuaded to pump more oil? Over lunch they were presented with the IEA's study, which showed that, if they were really so concerned, the solution was in their own hands. There were plenty of ways in which the West could moderate oil demand, thus bringing prices back under control. But they showed absolutely no interest! Sadly, perhaps they were right. The demand for oil is insatiable, and even what may seem to be quite reasonable and moderate measures to manage it may be politically unsaleable.

The long-term options include switching to more efficient, or electric, vehicles. But here are the IEA's options for reducing oil demand in a hurry, listed in descending order of impact on oil demand:

- Comprehensive and managed programmes of carpooling, including designated motorway lanes, park and ride lots.
- Driving restrictions based on odd/even licence plates.
- Reduce speed limits.
- Free public transport.
- Telecommunications drive, including working from home and teleconferencing.
- Ecodriving. Efforts to change driver behaviour.
- Reduction in public transport fares and increase in services.
- More bus priority lanes on roads.

---

[x] International Energy Agency, *Saving Oil in a Hurry*, 2005.

Most of these measures are more suitable for some countries than others. The IEA estimates the savings from the more effective options in the range of 2–5% of oil demand.

Some of the measures, such as driving restrictions or comprehensive carpooling, are too coercive to be brought in except in real emergencies. Others, such as free and extended public transport, may be too expensive. But most of the others, adopted in a moderate form, seem like common sense. Many governments are already enhancing the use of public transport, and encouraging ecodriving and various forms of flexible working.

6. Conclusion

Energy supply crises, both domestic and international, have a nasty habit of appearing suddenly, out of the blue. It pays to be prepared. All such emergencies are different and detailed plans may not fit the specific crisis that arises. As Field Marshal Moltke said, no plan survives contact with the enemy. But it is essential to set up the basic framework of communications that will be required and also to create options that can be put into place very quickly if and when needed.

Chapter 4

Energy and Human Welfare

1. The Revolution in Living Standards

The industrial revolution that began in England more than 200 years ago is spreading around the world. In historic times, the vast majority of people lived in poverty and only a tiny elite could aspire to living standards comparable to those that are widely enjoyed today in advanced societies. But that is changing, and changing with astonishing rapidity. According to the UN the proportion of people who live in extreme poverty, on less than $1.90 a day, has fallen by two thirds from 35% in 1990 to 10.9% in 2013, raising the life chances of more than a billion people. That improvement is reflected in the trends of life expectancy, education, and living standards. Rising prosperity is not just lifting poor people above the poverty line, it is also increasing the living standards of people higher up the scale, and creating large new middle class populations in the developing world.

This is quite simply the best thing that has ever happened. And the best thing that we can hope for is that this trend will continue. All this is obvious to people living in the developing world, but it may not be quite so obvious to those living in prosperous nations that have already been industrialised. Many boats have been lifted by the rising tide but, unfortunately, they have not included many average earners in rich countries. There are political consequences of this and, in the field of energy, it may limit the willingness of the rich countries to carry burdens for the benefit of the world as a whole. However in some respects, at least, the developed world deserves credit for having pursued trade and other policies that have supported the economic progress of the developing world.

The United Nations Development Programme (UNDP) publishes regular reports on human progress. Their 2013 edition gave a compelling account of how, in recent decades, industrialisation has transformed living standards, especially in China and India.

> The rise of the South has been unprecedented in its speed and scale. Never in history have the living conditions and prospects of so many people changed so dramatically and so fast. Great Britain, where the industrial revolution originated, took 150 years to double output per capita. The United States, which industrialised later, took 50 years. Both countries had a population below 10 million when they began to industrialise. In contrast, the current economic take-offs in China and India began with about 1 billion people in each country and doubled output per capita in less than 20 years, an economic force affecting a hundred times as many people as the Industrial revolution did. The rise of the South must be understood as the story of dramatic expansion of individual capabilities and sustained human development progress in the countries that are home to the vast majority for the world's people.[i]

If you read the Western media and, it must be said, some environmental literature, you could easily gain the impression that the main issue was not the growing health and prosperity of mankind but, rather, the pressure that the spread of industrialisation is putting on global resources. Of course, this is a serious issue and our options for coping with this are a major topic of this book. However, from the perspective of the developing world it may seem odd that the already developed countries, having exploited their own natural resources and scoured the world for resources to support their own development, now appear horrified about the environmental implications of the developing nations starting down the same course. Unless this is understood there is very little hope of reaching

---

[i] United Nations Development Programme (UNDP), *Human Development Report 2013, The Rise of the South.*

international agreement on how these environmental and resource issues are to be managed.

None of this is to deny the gravity of the environmental issues that are raised by mass industrialisation, especially in the field of energy and especially with regard to climate change, which is the topic of Chapter 5.

Energy has made a large and essential contribution to these advances in living standards. World energy demand increased by a staggering 55% between 1990 and 2014 and 80% of this increase was in developing nations, which almost doubled their demand. The combined coal consumption of India and China has risen from 1.2 billion tonnes in 1990 to more than 3.4 billion tonnes in 2014, accounting for 60% of world coal consumption and 25% of global carbon emissions.[ii]

But there is a long way to go yet. More than a billion people still live in the countries categorised by the UN as having "Low Human Development". For these countries, the average life expectancy at birth is less than 60 years, GNP per capita is $1,400, and the infant mortality rate, at 73 per 1,000 live births, is more than 10 times greater than for the most developed nations. These tend to be the countries where the population is growing fastest, at about 2.3% p.a. on average. A further 3.5 billion people live in countries categorised as "Medium Human Development" with GNP per capita of less than $6,000.[iii]

According to the IEA, "In developing countries, access to affordable and reliable energy services is fundamental to reducing poverty and improving health, increasing productivity, enhancing competitiveness, and promoting economic growth".[iv]

2. The Crucial Role of Energy

In 2008, an academic paper published in Energy Policy looked in depth at the relationship between per capita energy consumption and human

---

[ii] International Energy Agency, *World Energy Outlook*, 2016.

[iii] UNDP, *Human Development Report*, 2014.

[iv] International Energy Agency, *Modern Energy for All, Why it Matters*, IEA website, http://www.worldenergyoutlook.org/resources/energydevelopment/modernenergyforallwhyit-matters/. Accessed May 2017.

development as measured by the UN Human Development (HDI) Index.[v] The factors included in HDI are life expectancy, years of schooling, and per capital income. The authors chart energy consumption against the HDI for 90 nations. This shows that for countries at a low and medium stage of human development there is a very strong dependence of human development on increased access to energy. For more developed nations this relationship continues, although it is less clear cut. However, some of the most developed nations of all have very high levels of energy consumptions that are not significantly associated with improvements in human development.

The authors conclude that, "it is striking how absolute the requirement for energy is and how much can be realized from relatively small increments of energy available at the lowest levels...it is also striking how much excess energy is spent by the energy advantaged nations for no real improvement in quality of life as measured by HDI".[vi]

Of the 7 billion inhabitants of the Earth today about 1.25 billion, or less than one person in five, lives in a developed, OECD nation. On average, each person living in the OECD countries consumes more than twice as much energy and more than twice as much electricity, and is responsible for more than twice the CO_2 emissions, as the world population as a whole. So there is still a huge disparity. To put this another way, if the rest of the world was to catch up with the OECD by replicating its industrial systems then the world would more than double its energy consumption and carbon emissions.

Business as usual in the energy sector, in the absence of a low carbon transformation, does indeed represent a serious threat to human welfare, and to the environment more widely. Chapter 5, on climate change, is devoted to this issue. To protect mankind from dangerous climate change carbon emissions from energy, need to peak almost immediately and then decline sharply to half the current level by 2050.

---

[v] Martinez, D.M. and Ebenhack, B.W., Understanding the role of energy consumption in human development through the use of saturation phenomena. *Energy Policy* **36**(4), 2008, pp. 1430–1435.
[vi] Martinez, D.M. *et al.*, *Ibid.*

The original climate treaty of 1992 recognised that, "social and economic development and poverty eradication are the first and overriding priorities of developing nations". This is not to downplay the importance of urgent measures to mitigate climate change. But our concern for the impact of climate change on developing nations in 50 or 100 years' time must be matched by willingness to address today's problems of acute poverty and deprivation associated with lack of modern energy supply. This is increasingly being referred to as "energy justice".

The United Nations has recognised that,[vii] "energy is central to nearly every major challenge and opportunity that the world faces today. Be it jobs, security, climate change, food production, or increasing income, access to energy for all is essential". However, also according to the UN, today one in five people lacks access to modern energy and 3 billion people rely on wood, coal, charcoal, or animal waste for cooking and heating.

So energy is central to the prosperity of mankind as well as to the future of our environment. These are two sides of the same coin. It's an obvious point, but not always as well recognised as it should be.

Having reviewed the general contribution of energy to rising living standards, we shall now discuss more specific areas in which there is a link between energy and health. These are, local outdoor air pollution, the problem of indoor air pollution from traditional fuels, and the specific importance of electricity supply. But first, a brief excursion on the topic of underground coal mining, which represents the most important occupational hazard connected with energy.

3. Health Hazards of Coal Mining

There are occupational risks to health associated with many energy, and of course other, industries. But in terms of the sheer numbers of people involved, the coal mining industry, in deep underground mines, stands out. According to a Lancet study, up to 12% of coalminers develop one of several potentially fatal diseases; pneumoconiosis, progressive massive

---

[vii] *UN Sustainable Development Goals* adopted on 25 September 2015. Accessed at http://www.un.org/sustainabledevelopment/sustainable-development-goals/ on 6 April 2016.

fibrosis, emphysema, chronic bronchitis, and accelerated loss of lung function. China accounts for nearly half the world's coal production and probably more than half the occupational health impact.

Nearly 2.7 million workers are exposed to dust in Chinese coal mines, and dust concentrations in most mines far exceed regulatory standards.[viii] As a result 57,000 coal miners contract lung disease, or pneumoconiosis each year. By the end of 2007, there were a total of 312,000 cases. More than 6,000 workers die from pneumoconiosis, also known as black lung, every year. These figures do not include local coal mine enterprises in villages and towns, where safety rules may be very lax, so they may significantly under state the total. About half this number of coal miners die from mining accidents each year. The Chinese government is driving to improve safety in coal mines and so, hopefully, these statistics are improving.

It should be said that UK mines also, until the 1970's, suffered from inadequate dust control and the UK government still has a very large programme to compensate ex-miners with pneumoconiosis.

These hazards are not inevitable and modern deep coal mining in well managed mines is a great deal safer than it used to be. A big danger has always been the risk of explosion caused by the build-up of methane in the atmosphere. The famous Davey Safety Lamp, invented in 1815, was intended to prevent the methane from being ignited by miners' lights. It remains vital to ensure adequate ventilation of mines to avoid concentrations of methane in the first place. Some mines now extract methane from the coal in advance of mining, selling the methane, which is natural gas, as a valuable product.

The avoidance of lung disease is partly a matter of spraying water on mine workings to lay the dust and partly the wearing of masks by mineworkers, though this can be unpopular in rather a tough working environment.

The other major hazard is rock collapse, now usually controlled by driving long bolts into the roofs of mine "roadways" to hold them together.

[viii] People's Daily Online, *57,000 Chinese coal miners suffer from lung disease annually*, 11 November 2010.

One still, occasionally, hears of terrible mining disasters in different part of the world but these events should be avoidable today.

4. Outdoor Air Pollution

Local outdoor air pollution from fossil fuels is a serious health hazard. Almost all major cities are affected. The health effects of electricity generation were assessed by a paper in the Lancet in 2007.[ix] This showed that coal and lignite, or brown coal, had by far the greatest impact. Applying the fatality rate estimated in this paper to European coal generation rates in 2011 implies that they were responsible for over 20,000 deaths. In developing countries the rate of death may be considerably higher, due to less effective regulation. The same study gave an estimated death rate from a normal coal-fired plant meeting Chinese environmental standards that would imply some 140,000 fatalities in China due to coal generation in 2011. Applying the European estimate to OECD countries and this Chinese rate to non-OECD countries would give a global rate of death of just over half a million p.a. for the world as a whole. It must be acknowledged that standards have probably improved significantly since these estimates were made. But there is no doubt that coal power generation continues to represent a significant health hazard. Gas is much cleaner than coal both to produce and to burn. The health effects of gas power generation are more than an order of magnitude lower than for coal.[x]

According to the World Health Organisation (WHO),[xi] outdoor air pollution from all sources is estimated to have caused 3.7 million premature deaths worldwide in 2012. The causes include strokes, heart disease, lung cancer, and both chronic and acute respiratory diseases, including asthma. The sources of this pollution are not confined to energy but fossil fuels used for home heating, vehicles, and industry play an important part.

[ix] Markandya, L. and Wilkinson, P. (2007) Electricity Generation and Health, *The Lancet* **370**, 2007.

[x] Markandya, L. *et al.*, *Ibid.*

[xi] World Health Organisation, *Fact Sheet No. 313*, updated March 2014.

A report published by the IEA in 2016 gives an estimate of 3 million deaths in 2012, of which 35% were in China, and 21% in India.[xii]

Anyone who has visited Delhi or Beijing, or other industrial cities of developing nations, recently will realise the gravity of this problem. On bad days, visibility is closed down by yellow smog, the air tastes bad, one's throat feels rough, people start to wear face masks in the street. People worry, increasingly, about the health of children raised in the city. The global population is rapidly shifting to cities. In 1960, 34% of the world's population lived in cities but by 2014 the share was 54%. The shift was especially rapid in developing countries. It's no wonder that urban pollution, mainly from fossil fuels, is high on the agenda of many governments.

Governments around the world are working on this problem, but the solutions are not easy. The Chinese are working to move coal fired industry and electricity generation away from the Beijing region. But vehicles are also a big part of the problem. Reducing the share of diesel powered vehicles and managing traffic to avoid big jams can also help. But the air in cities will not really be clean until we have electric or hydrogen driven vehicles and non-fossil options for heating and cooling.

5. Indoor Air Pollution

Now we are turning to the equally serious problem of indoor air pollution, mainly the result of traditional cooking and heating methods.

The IEA estimate that more than 2.6 billion people, or 38% of the global population, rely on the traditional use of biomass for cooking. The situation is getting worse in sub-Saharan countries, but the largest numbers are in Asia. In India more than 800 million people, or about two thirds of the population rely on traditional biomass for cooking. In China about one-third of the population still relies on biomass, or over 400 million people.

Another study for the *Lancet* found that "household air pollution from solid fuels" was the third leading risk factor, for global disease in 2010,

---

[xii] International Energy Agency, *Energy and Air Pollution*, 2016.

coming only after high blood pressure and smoking in its impact.[xiii] The study estimated that this pollution accounted for 3.5 million deaths in 2010 and 4.3% of global years lost or lived with disability. In this study too, the main concentrations are in South Asia and sub-Saharan Africa. The main consequences are cardiovascular and circulatory diseases, chronic respiratory diseases, and lower respiratory and other common infections.

The main means of addressing this huge health burden are to enable households to switch from the most polluting fuels, traditional biomass and kerosene, to cleaner fuels such as Liquid Petroleum Gas (LPG) gas, or electricity, and to install cleaner, better ventilated cooking stoves.

There are a number of initiatives taking place to achieve these changes, for instance the Global Alliance for Clean Cookstoves is an international partnership developed by Shell Foundation, the UN Foundation and the US State Department. Established in 2010, its stated objective is to "save lives, improve livelihoods, empower women, and protect the environment by creating a thriving global market for clean and efficient household cooking solutions". The Alliance's target is that 100 million households will adopt clean and efficient cookstoves and fuels by 2020. They work with a network of public, private and non-profit partners to accelerate the production, deployment, and use of clean and efficient cookstoves and fuels in developing countries, such as Bangladesh, China, Ghana, Kenya, Nigeria and Uganda.

6. Access to Modern Energy

Without modern energy for heating, lighting, refrigeration, communications, and other purposes, human opportunities are diminished. The IEA have estimated[xiv] that 1.3 billion people, or 18% of the world population, did not have electricity supply in 2011. Most of these people are in sub-Saharan Africa, India, Pakistan, or Indonesia, with some in Latin America

---

[xiii] Lim, S. *et al.*, A Comparative Risk Assessment of the Burden of Disease and Injury Attributable to 67 Risk Factor Clusters in 21 Regions. 2012 *The Lancet* **380**, pp. 2224–2260.
[xiv] International Energy Agency, *World Energy Outlook*, 2013, Chapter 2 section on *Modern Energy for All*.

and the Middle East. After concerted efforts over many years, China and Brazil are close to achieving universal access. Recent improvements, from 2010 to 2011 have been modest, with the situation in some regions continuing to deteriorate. The problem is increasingly concentrated in rural locations and in sub-Saharan Africa.

As the *Lancet* has noted, "access to electricity is a prerequisite for the achievement of health, and lack of access to it remains one of the principal barriers to the fulfilment of human potential and wellbeing".[xv] A 2001 World Bank study[xvi] of more than 60 poor countries found that in urban areas linking households to electricity was the only key factor that reduced both infant mortality and under 5 years old mortality and that this effect was large, significant, and independent of incomes.

According to the UN,[xvii] access to modern energy is also a gender issue:

> without access to modern energy, women and girls spend most of their day performing basic subsistence tasks…of the estimated two million annual deaths attributable to indoor air pollution generated by fuels such as coal, wood, charcoal and dung, 85% are women and children who die from cancer, acute respiratory infections, and lung disease….in fact illnesses from indoor pollution result in more deaths of women and children annually that HIV/AIDS, malaria, tuberculosis and malnutrition combined.

If the world is willing to make an effort to address the energy related problems of developing nations, it would seem reasonable for access to modern energy to be the first priority.

Energy access is primarily a rural problem. Extending electricity supply into relatively remote regions, whether through grid extension or standalone supply, can be a big problem for a developing nation with

---

[xv] Markandya, A. and Wilkinson, P. *Electricity Generation and Health, The Lancet* **370**, 2007.

[xvi] Wang, L. *Health Outcomes in Poor Countries and Policy Options: Empirical Findings from Demographic and Health Surveys.* WPS2831. Washington, DC: The World Bank 2001.

[xvii] United Nations Industrial Development Organisation (UNIDO) and UN Women *Sustainable Energy for All: The Gender Dimension*, 2013.

rapidly growing energy demands. Help through International Financial Institutions or development aids can make an important contribution. Developing nations also need to provide a stable regulatory framework that encourages private investment.

One of the UN's Sustainable Development Goals is to, "ensure access to affordable, reliable, sustainable and modern energy for…".[xviii] In 2011, UN Secretary General Ban Ki-moon launched his Sustainable Energy for All (S4All) initiative, citing lack of access to electricity and to modern cooking and heating as, "a major barrier to eradicating poverty and building prosperity." Developing countries as well as many businesses, aid organisations, and NGOs have joined the initiative. Ensuring universal access to affordable, reliable, modern energy services is one of the UN's Sustainable Development Goals. The IEA has drawn up an "Energy for All" scenario which would meet SE4Alls target of universal access to electricity and clean cooking by 2030. The increased electricity supply raises total energy demand by less than 1%. Three quarters is assumed to come from renewables. There is also additional demand for LPG. The total increase in CO_2 emissions is only 0.7%. On this basis, the environmental penalty seems modest.

7. Building Modern Economies

It is natural and inevitable that the population of developing nations should aspire not just to basic energy access but to energy services to support a standard of living that approximates to the living standards enjoyed in the West. There is still a huge unsatisfied demand. The challenge that we face is to meet this demand at acceptable cost and to moderate and then reduce global carbon emissions at the same time. It's quite a tall order. And it's at the heart of the energy conundrum.

Economic development means building a national infrastructure for clean water and sanitation, decent homes, schools, hospitals, roads, factories, and power stations. All of this requires a lot of energy and energy-intensive materials such as cement and steel. It would be great if the developing world could ignore the example of the West and find an alternative

---

[xviii] *UN Sustainable Development Goals*, UN website.

low energy development route. Technological progress, discussed in Chapter 6, may yet open the door for this. Developing nations would like nothing more than to "leapfrog" the West by moving directly to advanced technologies that are cheaper, more efficient and cleaner than conventional energy. But, today fossil fuels are still amongst the lowest cost and most accessible energy sources in many parts of the developing world and large investment is required to make the transition. Bearing in mind the slow pace at which much richer developed countries are trying to curb their profligate energy use and switch to low carbon sources there is a limit as to what can be expected of the developing world. It is reasonable that less developed nations have said that they must rely on financial and other help from the developed world to support their low carbon transition.

In the following sections, we turn to the question of how some major developing regions are trying to meet their energy needs while managing the environmental consequences.

8. China

China is the most populous nation on earth with 20% of the world's population. Economically, it is also one of the most dynamic. In terms of purchasing power parity China now accounts for 13% of global GDP. China's rapid and sustained growth over several decades has been concentrated in heavy industry and construction, with the result that China now accounts for 24% of global industrial energy consumption.[xix]

By sustaining an astonishing growth rate of around 10% p.a. over the past decade, China has more than doubled its GDP and measured in purchasing power parity, its economy is now of a similar size to that of the US. China has recently overtaken the US to become the world's largest energy user.

China's rapid economic development has generated "the most rapid decline in absolute poverty ever witnessed" according to the UNDP. Hundreds of millions of people have benefited. But China remains a developing country.

[xix] This section is partly based on Hirst, N. *et al.*, *An Assessment of China's Carbon Intensity Target*, Grantham Institute Report GR1, 2011.

China has achieved the goal of halving the number of people in extreme poverty by 2015 set by the UN, in 2000, as one of its millennium development targets. China's GDP is about $13,000 per head at Purchasing Power Parity, which is still low by OECD standards. But there is a wide variation between provinces. Guangdong, a province of nearly 100 million people, has GDP per head that is much higher. There are major centres of population away from the east coast where GDP per head is much less. Remaining rural poverty is now concentrated in remote regions, but China faces a new challenge in the form of urban poverty related to migration to the cities.

One consequence of China's rapid economic growth has been that it has moved from being largely self-sufficient in energy as recently as 1980, to importing about half its oil needs today. This oil import dependency is set to increase further with the rapid growth of the private car market. China is now the world's largest market for new cars and China alone could account for more than half the increase in global oil demand in the coming decades. China is currently only 5% dependent on imports for its gas supply, but this is expected to increase sharply in future years.

As mentioned in the previous chapter, coal is the most accessible, secure, and, in many locations, the lowest cost source of energy in China. In 2014, coal fuelled more than 70% of China's power generation and contributed 65% of total energy supply.[xx] China is largely self-sufficient in coal, and has proven reserves equivalent to 45 years of production at current rates. However, it is possible that China's coal consumption has now peaked, as the rate of economic growth has moderated, and China seeks a cleaner and less energy-intensive economic balance. Nevertheless, high dependency on coal is expected to continue for some decades. This, of course, represents a problem from the point of view of climate change as well as local pollution.

China is clearly aware of the need to improve its energy efficiency and reduce its reliance on fossil fuels, which brings not only potential economic benefits such as security of energy supply and a reduced oil import bill, but local and global environmental benefits as well. Through its recent Five Year Plans, the government has been pursuing aggressive

---

[xx] International Energy Agency, *World Energy Outlook*, 2016.

energy efficiency policies. In the 11th Five Year Plan, for 2006–2010, China set an ambitious target to reduce energy intensity (energy consumption per unit of GDP) by 20% on 2005 levels. In the event, a 19.1% reduction was achieved. But radical measures, including power cuts and plant closures, were reported during 2010 as a part of China's efforts to meet this target.

In announcing the next 12th Five Year Plan (2011–2015) the Chinese government clearly signalled a focus on sustainability and the environment. Then Prime Minister Wen Jiabao said, "We must not any longer sacrifice the environment for the sake of rapid growth and reckless roll-outs, as that would result in unsustainable growth featuring industrial overcapacity and intensive resource consumption". The Plan contained an overall economic growth projection of 7% p.a., significantly lower than actual growth in previous years. It also contained a number of energy and emissions targets including an energy intensity reduction of 16% and carbon intensity reduction of 17%, on 2010 levels. At the same time, the Plan identified seven strategic emerging industries (SEIs) as being critical to China's economic development, including electric vehicles, energy efficient products and renewable energy. Investment in these industries was approximately $1.5 trillion over the course of the Plan. The Plan also included major increases in non-fossil energy, including a four-fold growth in nuclear power to 40 GW, 63 GW of new hydroelectric capacity, 48 GW of new wind capacity and 5 GW of solar capacity by 2015. A Gigawatt is a billion watts. For comparison, total UK generating capacity is about 60 GW.

The 13th Five Year Plan for 2015–2020 will continue this process. Energy intensity is set to decline by 15% over the period and carbon intensity by 18%. Given that there has been over-achievement of the targets for 2011–2015, this means that China is well on course to meet its target, set for the Copenhagen climate summit in 2009, of reducing carbon intensity by 40–45% between 2005 and 2020.

The new plan sets targets for an increase in the share of non-fossil energy from 12% in 2015 to 15% in 2020 and 20% in 2030. Given that China already has the world's largest renewables capacity, this implies massive new investment. The targets for hydropower, nuclear, wind and solar capacity are 340 GW, 58 GW, 200 GW, and 100 GW respectively.[xxi] Quite possibly these targets will be exceeded, as renewables targets have been in the past. Natural gas consumption is expected to increase to about 10% of total energy consumption.

However, there are areas of concern. Currently, and even though existing coal power stations are operating well below capacity, Chinese power companies are planning to invest in 210 GW of new coal capacity. The government in Beijing will need to find ways to curb these ambitions, something that may be consistent with its plans to reform heavy industrial state owned enterprises more generally. China's coal consumption may have peaked, but even this is not certain, and the decline may be slow. The IEA are projecting that even by 2040, when carbon emissions will need to be much lower than today, coal could still be much the largest source of electric power in China.

Another concern is that China is not making full use of the wind capacity already installed because of weaknesses in the grid and its regulation and because there is very little storage capacity. Strengthening the grid to better integrate variable renewables will be a challenge for the future for China, as it is in many parts of the world.

Today, China's power grids do not call on individual power stations in order of their cost efficiency but, rather, each station is given a percentage of its maximum capacity that it will be allowed to deliver each year. Regional governments and utilities continue to build additional fossil fuel and renewable generation even though, in many regions, there is already excess capacity because each new station receives its quota and the construction contributes to economic growth and employment. It's an inefficient system and it means that coal power sometimes runs ahead of wind or solar power with much lower running costs and emissions. Beijing has announced major reforms of this system to make it much more competitive and market driven, but this will no doubt face considerable institutional resistance.

Chinese industry pays about twice as much for its energy as industry in the US. This is one of the big drivers for reform. No doubt the difference is due, in part, to America's cheap gas and coal resources. But the less competitive Chinese energy market is probably also a factor.

[xxi] China Daily, *China issues five-year plan to reduce greenhouse gas emissions*, 5 November 2016.

China's role in climate negotiations will be discussed in Chapter 5. China took a leading role, with the US, in promoting the success of the Paris climate summit of November 2015 and China, now, definitely shares in the leadership of the Paris climate process. Following President Trump's announcement of American withdrawal, the leadership now rests with China and Europe. Specifically, China made a commitment that its carbon-dioxide emissions will peak by 2030, and sooner if possible. This stance reflects, in part, China's ambition to rebalance its economy to a cleaner and less energy-intensive model. Probably, in practice, China will do better than this. Many commentators think that coal consumption has already peaked and that total greenhouse gas emissions will peak well before 2030. The fact that these targets are, to some extent, now the subject of international negotiation may have made China rather cautious in the reductions that it offers in the Five Year Plan.

It is worth reviewing the options available to China as it rebalances its energy sector. These can be considered under four broad headings, electricity generation, industry, buildings, and transport.

In electricity generation the main options are increasing the capacity of wind, solar, hydro, and nuclear power and biomass, switching from coal to natural gas, and raising the efficiency of the coal power fleet. Because the need is so great, China is pursuing all these options.

China has some of the most efficient industrial plants in the world. But there is also a big tail of older, generally smaller, and less efficient plants. China has huge potential to moderate the growth of carbon emissions, especially through the modernisation of iron and steel, cement, and chemical plants.

China has been going through a phase of very rapid growth based on high levels of basic manufacturing and infrastructure development. As a result, heavy industry and construction account for a much larger share of the economy than would be usual for a developed nation. One of the aims of the government is, over a period, to rebalance the economy so that lighter, higher technology and higher value added industries, and the service sector, account for a larger share, driven by increased domestic consumption. This should help to limit the growth of China's energy demand and carbon emissions and reduce local pollution. Of course, it is possible that some of the more polluting basic export industries will move to elsewhere in the developing world where labour costs are still low.

Rapid economic growth and increasing living standards in China have been accompanied by a boom in construction. As a result, China accounts for about half of the world's cement manufacture. The energy efficiency of this huge building stock will have a significant impact on Chinese and global emissions that, once construction has taken place, will largely be locked in for the life of the buildings. That is why Chinese building standards are so important, especially in the northern, colder parts of the country. So tightening up the regulation of the efficiency of residential and commercial buildings, and its enforcement, will be essential. The other important elements of a low carbon strategy for buildings are more efficient lighting and appliances and increased use of district heating.

The final sector is transport. At one stage, China had great hopes of "leapfrogging" the West by adopting electric vehicles. But China is now the world's largest market for conventional, internal combustion, cars and vans. Today's efforts are concentrated on extending public transport, improving vehicle efficiency, and increasing the share of low carbon, biological fuels such as bioethanol and biodiesel. Electric vehicles will no doubt have a big impact in the future. They are starting from a very low market share today, but China is starting to set tough new quotas.

China is believed to have the world's largest shale gas reserves, exceeding those of the US. However, the geology is more difficult and China lacks America's highly developed oil and gas culture. Increased gas supply could help China to reduce its dependence of coal and also ease the crisis of air pollution in cities. Some would say that this is merely replacing one polluting fossil fuel with another. It is true that shale gas development, if not properly regulated and managed, can yield worrying emissions of methane, an intensive greenhouse gas. But gas yields only about half the climate emissions of coal in combustion, and shale gas has already significantly reduced climate emissions in the US. A pragmatic view would be that this is a major opportunity to reduce China's carbon emissions and air pollution in the medium term that we cannot afford to miss.

China may also be moving, more generally, from an inward looking managed energy sector to one that is more outward looking and market based. How this is to be achieved, and the rate of change, are hotly debated in Beijing today. The outcome will be of great importance for world energy. Probably a more market-based approach would be good for energy efficiency, and for shale gas development, and would facilitate the ending of subsidies for fossil fuels. More competitive markets could be good for renewables as they become increasingly competitive.

The Chinese government is adopting cap and trade mechanisms in selected regions. Six municipalities and provinces including Beijing, Chongqing, Shanghai, Tianjin, Hubei, and Guangdong are already starting a carbon trading pilot scheme. These are encouraging indications of the possible direction of future policy. According to the 13th Five Year Plan there is to be a national carbon trading market operating by 2020. This would be the world's largest such scheme by far. The question of how it might eventually relate to other trading schemes in Europe or North America is of great interest for the future of world carbon regulation.

China is now at an intermediate stage of development. A period of astonishing growth based largely on manufacturing for export, infrastructure investment, and heavy industry is now moderating. Unlike many other developing countries, China is not short of electricity. Its existing coal power stations are operating well below capacity and, for the time being, the central government is trying to rein in new investment. China is also, now, aiming to limit new mining operations as coal prices fall. As the most basic needs of the population are increasingly being met, attention is turning to the environmental consequences at both local and global levels.

China today can be seen to be conducting a huge experiment in how growth can continue, at more moderate rates, and living standards continue to rise, but in a way that is much less energy and carbon intensive. This strategy may be tested by some Chinese provinces whose economic development lags well behind that of the wealthier east coast regions. China has started from an exceptionally high degree of dependence on energy-intensive industry. If China's strategy is successful it will provide an important example for other developing nations seeking to reconcile economic growth with protection of the environment.

China's success, or otherwise, in rebalancing its economy and controlling carbon emissions will be one of the biggest factors in climate change. Is there anything the West can do to facilitate this? There are many government and NGO programmes. These include cooperation on, energy market reforms, on the design of efficient buildings and cities and on appropriate regulation and enforcement. Various forms of business cooperation are contributing to the development and deployment of renewables, where China already leads the world, and more efficient power stations and factories. International organisations, such as the IEA, can ensure that China is part of the international debate on sustainable energy policies.

9. India

India is the second most populous nation on earth with a population of 1.3 billion in 2015 representing about 17% of the world's population. India's economy has been growing, on average, between 6% and 7% annually in recent decades and at 11% between 2004 and 2014. Energy demand increased by more than 150% between 1990 and 2014, making India the third largest energy consumer in the world, after China and the US. CO_2 emissions nearly tripled, to over 2 billion tonnes, about 6% of the global total.[xxii]

The service sector plays a much bigger role in India's development than it does in China's. 68% of India's economic growth in the period 2001–2008 came from the service sector, whereas in China the figure was 42%. Conversely, 52% of China's growth was from industry, which accounted for only 27% of India's growth. Raising living standards in India will certainly require a large increase in energy supply. Nevertheless, India will probably never pursue quite such an energy-intensive pathway as China has done.

India's energy economy is largely coal based, like China's, with 76% of power generation derived from coal plants in 2016. India is currently the world's third largest coal producer, and is projected by the IEA to become the second largest, after China, by 2030. India has about 7% of the world's proven coal reserves, representing about 122 years of production at current rates.

[xxii] International Energy Agency, *World Energy Outlook*, 2016.

However, India's coal reserves are generally of poor quality and coal supply is constrained by environmental restrictions on access and by inefficient mining technologies. Also, India's coal reserves are concentrated in the northeast, and it is costly to transport domestic coal to the south and west of the country. For these reasons substantial coal imports, mainly from Australia and Indonesia, are likely to be needed to meet India's growing energy needs.

India depends on imports for more than three quarters of its oil supply. This means that the recent sharp fall in oil prices in 2014 was a major economic boost. Rather than allow the fall in oil prices to benefit consumers directly, and perhaps increase oil demand, the government has chosen to increase oil taxes and take the benefit in government revenues. Gas contributes a relatively modest 6% of India's energy demand and a growing share of this is now imported.

India's Expert Committee on Integrated Energy Policy, reporting in 2006, recognised that India faced formidable challenges in meeting its energy needs. Their vision required that India "pursues all available fuel options and forms of energy, both conventional and non-conventional. Further, India must seek to expand its energy resource base and seek new and emerging energy sources. Finally and most importantly, India must pursue technologies that maximise energy efficiency."

In June 2008, the prime minister of India released India's first National Action Plan on Climate Change (NAPCC), which identified eight core national missions running to 2017. These included a national solar mission and a mission for enhanced energy efficiency. The Perform, Achieve, and Trade (PAT) scheme, now in the process of implementation, is a key element of the energy efficiency mission and will focus on more than 700 of the most energy-intensive installations in India.

Nearly 30% of India's population still lives in "severe poverty"[xxiii] and, therefore, the government's first priority is economic growth, development, and poverty alleviation. Reform of the power sector, increasing the quality and quantity of coal supply, and investing in new and efficient coal power stations, have all been major government priorities. Curbing the

---

[xxiii] UNDP, *Human Development Report*, 2014.

growth of carbon emissions is certainly an objective but it will only be pursued through policies that also contribute to economic development. As the Indian government points out, India's carbon dioxide emissions per person are less than one fifth of those of the developed world[xxiv] so it is hard to expect India to take a leading role.

Nevertheless, India is the country with the largest projected growth in energy related carbon dioxide emissions over the coming decades. Due to India's outstanding growth prospects, the increase in India's CO_2 emissions between 2014 and 2040 are projected by the IEA, in their baseline case, to equal about three quarters of the net global increase.[xxv] So India is a crucial part of the equation if we are to mitigate climate change successfully.

Many of the carbon reducing options for India in the medium-term are not very different from those for China. In the power sector they include solar and wind, hydro and nuclear, as well as modernisation of coal power plant, and switching to gas. India has immense potential for solar power, and the government has declared its "mission" that solar power should replace coal as India's main source of energy.

In the industrial sector they include modernisation of industry, including steel and cement, for instance through the PAT scheme. In the transport sector, vehicle efficiency improvements and better public transport are key.

In an effort to reduce air pollution in large cities, India has introduced public transport that runs on compressed natural gas, a world leading initiative that has the potential for further development. There is less need for domestic heating in India than in China, but gains in the building sector are possible through efficient lighting and more efficient appliances. The efficiency of air conditioning is becoming increasingly important as affluent families demand cooler living conditions.

India suffers big problems in electricity distribution and supply. The system still suffers, in many states, from regulated prices that are below cost and from supply that is unpaid for. As a result the distribution system

---

[xxiv] International Energy Agency, *Key World Energy Statistics*, 2014.
[xxv] International Energy Agency, *World Energy Outlook*, 2016.

is weak, sometimes unreliable, and unable to invest in new infrastructure. Reform is desperately needed to open the door for the investment that can make reliable power more accessible to some 200 million people who are still without it.

In the past India has also suffered chronic difficulties in commissioning sufficient new generating capacity, but the situation has improved dramatically in recent years, especially with increased private sector participation and with the addition of large volumes of renewables capacity. In the years 2022–2027 India expects to add 44 GW of coal, 12 GW of hydro, 40 GW of wind, 50 GW of solar, and 7 GW of biomass electricity. Because of the intermittent nature of wind and solar, coal is still the biggest contributor in terms of the electricity that the new capacity will supply. Nevertheless, this represents a major renewables revolution for India. Because of the rapid growth of renewables, and some 50 GW of new coal capacity that is already in the pipeline, the Central Electricity Authority have said that they do not expect to need any further additions to coal capacity in the years 2022–2027.[xxvi]

Nevertheless, the Indian government expects that, "in order to secure reliable, adequate and affordable supply of electricity coal will continue to dominate power generation in the future".[xxvii] The renovation and modernisation of coal plant as well as the construction of new and efficient large "super-critical" stations are an important part of the government's low carbon strategy. The government is still reported to be aiming to double coal production capacity to 1.5 billion tons p.a. by 2020.[xxviii]

In its Nationally Determined Contribution to the 2015 Paris climate summit, India set itself the target of reducing the emissions intensity of its GDP by 30–35% by 2030 from 2005. India also aims that, "with the help of transfer of technology and low-cost international finance", 40% of its generating capacity will be non-fossil based by 2030.

[xxvi] Central Electricity Authority of India, *Draft National Electricity Plan*, December 2016.
[xxvii] Government of India, *India's Intended Nationally Determined Contribution: Working Towards Climate Justice*, 2015.
[xxviii] Asian Review, *Modi looks to double coal production by 2020,* April 2015.

Even taking account of this National Contribution, the IEA is project-ing that India's carbon dioxide emissions will more than double to 4.5 billion tonnes in 2035, much the largest national increase.[xxix]

This means that there is all to play for in the future of Indian power. The IEA have been projecting, in their 2015 Special Report on India, that India's power system will need to grow by three and a half times by 2040, increasing at almost 5% a year to support rapid economic growth. The share of coal in electricity generation falls from nearly 75% today to 57%. Nevertheless, the absolute amount of coal generation increases two and a half times so that India contributes almost half of the world's increase in coal generation. In spite of this rapid growth in supply, energy demand per capita in India in 2040 is still 40% below the world average.

Is this outlook too pessimistic? Certainly, India's appetite for renewa-bles continues to grow as cost falls. But much may also depend on the regulatory framework and the willingness of international investors, gov-ernment as well as private sector, to invest in India's low carbon energy revolution.

In its Intended Nationally Determined Contribution to the Paris cli-mate summit, India has laid down the gauntlet,

> India has shown its commitment to combat climate change and these actions are indeed important contributions to the global effort. However, our efforts to avoid emissions during our development process are also tied to the availability and level of international financing and technol-ogy transfer since India still faces complex developmental challenges. The critical issue for developing nations is the gap between their equita-ble share of the global carbon space and the actual share of carbon space that will be accessible to them. The transfer of appropriate technologies and provision of adequate finance will have to be a determined contribu-tion of the developed countries, which will further enable the developing countries to accomplish and even enhance their efforts. It is expected that developed countries would recognize that without means of imple-mentation and adequate resources, the global vision is but a vacant dream.

[xxix] International Energy Agency, *World Energy Outlook*, 2016.

They have a point. The response to this should be an important part of the carbon reduction efforts of the developed world.

10. A Digression — Australia's Export of Coal to India

What are the ethics of Australia exporting coal to India? This is a question being debated, in May 2017, by the government of the Australian state of Queensland. An Indian owned company, Adani, is proposing to invest $12 billion in what would be one of the world's biggest coal mines. The project has been criticised by environmentalists and the Government has yet to agree on whether to offer tax concessions. Arguably this coal is needed to support economic progress in India. The coal is of better quality and would probably be burnt more efficiently, with somewhat lower carbon emissions, than Indian coal. But it carries a huge environmental penalty in terms of carbon emissions.[xxx] Australia derived 63% of its own electricity from coal in 2014–2015, so is there an element of hypocrisy in trying to deny access to the same coal to India? One could perhaps consider this in the context of what efforts the Australian government is willing to make to reduce its own coal consumption and to help India onto a low carbon pathway, for instance by supporting energy efficiency or renewables projects or the deployment of carbon capture and storage. Without a contribution of this kind the investment is suspect.

In fairness to Australia it is worth noting that even governments that are taking a lead in carbon reduction domestically can be reluctant to allow environmental policy to affect export opportunities. Norway is a major exporter of oil and gas. Denmark is also an oil exporter, though on a diminishing scale. Germany is, by far, Europe's biggest exporter of carbon emitting cars. But coal is arguably different because it is the most carbon intensive of the fossil fuels and because, at least in principle, alternatives are available.

[xxx] Financial Times, *Australian Coal Mine Shelved in Subsidy Battle*, 23 May 2017.

11. Energy Justice

The concept of energy justice was explained to me by a senior Indian politician at a conference that I attended last year in Singapore. It goes like this. The Western world raised its living standards many generations back through the original industrial revolution. That revolution polluted the atmosphere and initiated the problem of climate change that we are now all facing. Now it is the turn of the developing world including India, and other countries where desperate poverty persists. Carbon dioxide emissions per person in India are still only a fraction of those in the developed world, about 17%. India is well disposed to climate mitigation and the reduction of greenhouse gas emissions, and to the introduction of advanced energy technology. But India's priority is the relief of poverty and raising of living standards. International help to achieve this is welcome. But it is unjust to lay upon India the burden of taking any steps towards climate mitigation that are suboptimal from the point of view of poverty relief and the development of India's own economy. This is perfectly consistent with international programmes to assist India with its ambitious plans for photovoltaics. But it is not consistent, for example, with blaming India for building new power coal power stations so long as that remains a part of the most practical and economic means to increase power supply. If the West wants India to take a low carbon road it will have to provide the support necessary to make that the most efficient way for India to achieve its development strategy. Perhaps R & D to reduce the costs and increase the flexibility of low carbon systems can achieve the same result. To put this another way, support for economic development must go hand in hand with support for the low carbon transition.

12. Africa

Sub-Saharan Africa is the least developed region on Earth, with life expectancy at birth of less than 60 years and GDP per person of just over $4,000 p.a. measured in purchasing power.[xxxi] Of course this is not one

---

[xxxi] UNDP, *Human Development Report*, 2014.

nation but a diverse region with many different governments and political complications. It has the lowest HDI of any major region. 49% of the population still live in absolute poverty. This figure has declined from 56% in 1990, as a result of improvements in the rate of economic growth, but because of rapid population growth the absolute number of people living in poverty is still increasing.[xxxii]

A generally under-developed energy sector contributes to this low level of economic development and health. According to the IEA, sub-Saharan Africa has more people living without access to electricity than any other world region, more than 620 million people representing nearly half the global total. Nearly 730 million people in sub-Saharan Africa rely on the traditional use of solid biomass, wood, straw, charcoal, and dried animal dung for cooking, typically with inefficient stoves in poorly ventilated spaces.[xxxiii]

The whole of Africa's CO_2 emissions per person are even lower than India's. With 15% of the world's population, Africa has less than 9% of the world's energy and contributes only 3% of world CO_2 emissions. As with India, it is hard to expect Africa to take a leading role in carbon reduction. However, the expectation is that the continent's carbon dioxide emissions will increase by some 800 million tonnes p.a. by 2040, mainly the result of increased oil and gas consumption.

The irony is that Africa is rich in energy resources. Angola and Nigeria are major oil producers. Nigeria, Mozambique, Angola, and Tanzania have major gas potential. South Africa is one of the world's major coal producers, and there is large scale hydro potential, especially in central and southern Africa. There is also huge potential for renewables, including wind, solar, and geothermal energy in the East African Rift Valley. The vast solar potential of North Africa is being considered as a possible source of low carbon energy for Europe. The IEA have pointed out, however, that since 2000 two thirds of the investment in energy in sub-Saharan Africa has gone into exports and not to meet the region's acute energy needs. It is widely recognised that poor governance has held

---

[xxxii] International Energy Agency, *Africa Energy Outlook, World Energy Outlook*, Special Report, 2014.
[xxxiii] *Africa Energy Outlook, Ibid.*

back economic development in parts of Africa and the IEA's analysis shows a clear correlation between measures of the quality of governance and the level of investment.[xxxiv]

In 2010, I was asked by the World Bank to advise on the environmental implications of investing in a huge coal power station at Medupi in South Africa. The case is described in more detail in Chapter 2. The power was desperately needed, but the environmental consequences were dire. This nicely encapsulates the dilemma that African governments and international aid organisations face in addressing the region's energy challenges.

Africa's vast energy resources are generally differentiated among its regions and are concentrated in a few countries within the different regions. Northern Africa dominates the production of natural gas and this activity is concentrated in Libya and Algeria. The West African coastline has significant crude oil deposits concentrated in Nigeria and Angola. The energy resources of the Eastern African region are dominated by all the countries along the Rift Valley, as they all have significant reserves of geothermal energy. Hydropower dominates the energy sources of central Africa with much of this resource being in the Republic of Congo. 90% of coal deposits are in South Africa, which has substantial reserves.

The challenge is to develop these resources and build the associated infrastructure that is needed. Developing the power sector in the continent will require overcoming certain obstacles. Although South Africa itself has relatively low energy costs, due to its coal resource, in many African countries, due to lack of investment, the cost of producing power and the price to consumers are still among the highest in the world. The investment needs are enormous but both public and private funds available are far from adequate. Several power institutions in African countries are pursuing institutional reforms to unlock their potential but so far their overall impact is limited.

Since the 1950s, power plants installed in Africa have been an average of only 40 MW per 1 million inhabitants when in the same period the figure was 200 MW in Latin America and 800 MW in Asia. Unfortunately, electricity growth rates in the continent have been very small. As a result

---

[xxxiv] *Africa Energy Outlook, Ibid.*

only about 35% of Africans have access to electricity and only about 24% below the Sahara. Electricity consumption for Africa outside of South Africa is only about one fifth of the level in China and less than one tenth of the average level in the developed OECD countries.

Today, Africans use only a small fraction of the amount of energy, per person, that developed countries use. As a result, Africa's contribution to climate change is modest. But Africa is a continent with a rapidly growing population of more than a billion inhabitants. It has been held back, in some countries, by problems of government, but today there are encouraging signs that economic growth may be picking up. In sub-Saharan Africa especially, energy shortages are one of the main constraints on progress and massive new supplies will be needed.

There are many multilateral and bilateral programmes of support for African countries to develop their energy sectors. These are important to provide much needed energy and to take advantage of opportunities for low carbon power and improve energy efficiency. Because some African countries are lacking in comprehensive electric grids, there are opportunities for self-contained networks supplied by renewables to provide early access to power in rural areas.

To a much greater degree than India or China, Africa has proved to be rich in oil and gas and other natural resources. In a number of African countries this can provide the finance and energy needed to accelerate development. But Africa has also been vulnerable to the "resource curse" in which resource wealth is diverted or provokes civil unrest. Energy resources have been exploited with little benefit to national citizens.

It is important that international business exploiting African resources for export should not be complicit in these problems. They need to account transparently for all payments and to ensure that local people and local businesses share in the benefits of their projects. There has been a lot of diplomatic activity around the Transparency Initiative, which is the main project in this area. Governments and NGOs should redouble their efforts and be tough on the international energy industry, including state-owned companies.

Africa's vast energy resources have the potential to make a large contribution to economic development and to improvements in living standards on the continent. The international community has an obligation to facilitate that. It is not reasonable to expect Africa, with its low carbon emissions per person, to carry economic burdens in order to mitigate climate change. But as the costs of renewables decline, and national energy administrations look to adopt more efficient energy systems, there is growing scope for energy options that serve the cause of carbon reduction as well as economic development.

13. Financial Support from the Developed Nations

In the climate negotiations, the developed nations have said that, from 2020, they will commit $100 billion a year to assist the developing nations to control their greenhouse gas emissions and adapt to climate change. This is certainly a lot of money. But it is not disproportionate when one considers that the total cost of meeting the 2°C target is variously estimated in the range of $1,000 billion p.a. The developed nations must make good on this promise, because this is their sign of good faith in energy relations with the developing world. Otherwise, it is hard to see how they will win the cooperation of developing nations for a low carbon strategy.

Developed nations are insisting that commercial finance, as well as favourable loans from financial institutions, can count against that total and there is no commitment on what share of the funds will be government grants. Most of the government finance given today is through bilateral programmes, in which each donor country pursues its own agenda, and not through the institutions of the climate treaty where the management is shared with the developing countries themselves. And there is even doubt as to whether all this money is genuinely additional to existing development aid programmes.

Bilateral aid and commercial funding are very important. Most of the international funds for energy development will come from these sources. Also, there is certainly a big overlap between the objectives of development aid and climate support. The developed nations have to be careful with their taxpayers' money. Nevertheless, there is a need for greater clarity as to how much of the $100 billion p.a. will be additional government funding and what share is to be provided through the Green Climate Fund.

The greatest threat to the promised $100 billion p.a. today is President Trump's announcement of US withdrawal from the Paris climate agreement, since the US was due to make the largest single contribution. It remains to be seen whether others will pick up their share. This is an important issue because the developing nations undoubtedly regard this offer of support as integral to the whole climate deal.

14. Conclusion

The ultimate objective of energy policy is to promote human welfare. The achievements of industrialisation, based largely on fossil fuels, have been spectacular, especially in the last couple of decades. Now we need to provide modern energy to raise the living standards of the many people who are still living in poverty, to bring electricity to those who are without it, and to address the scourges of indoor and outdoor air pollution. There is a lot of work still to do in all these areas. We need to find ways to achieve these things in such a way as to limit global warming, the topic of the next chapter, and itself a major threat to human welfare in the future.

The IEA have shown that there is very little conflict between climate objectives and the need to provide basic energy access and safe cooking stoves, because the volume of energy required is fairly modest and renewables may be suitable for at least some of these needs. But the way developing countries meet their wider energy needs for economic growth and development certainly is critical for climate objectives.

It is the energy strategies of the major developing nations that will largely decide the outcome on global warming, for the simple reason that this is where most people live. These strategies are bound to focus on the pressing energy needs of their populations. Governments of the developed nations need to work with these developing nations, and support them financially, to find affordable strategies that meet their development needs as well as climate goals.

Chapter 5

Climate Change

1. Introduction

The knowledge that human activity is changing the atmosphere also changes our perceptions of the mutual dependence of mankind and of our relation to the planet. The earth seems smaller and more vulnerable. Our shared responsibility for protecting its extraordinary life-sustaining qualities has become palpable.

Of course, mutual dependence is not new. Human progress has always been collective. We live in a global village in which just about all aspects of life have a worldwide dimension. But climate change takes this a step further because of the very direct impact of greenhouse gas emissions anywhere on our shared environment and because of the risk of irreversible harm. So far we have managed without a world government, although the United Nations does provide a framework for cooperation between sovereign states. Now, the threat of climate change provides a unique test of mankind's ability to work together in the common interest.

This chapter will discuss the evidence for climate change and its likely consequences, the economic and ethical debate on how we should respond, and then the history of international mitigation efforts. Finally, it will review the costs of climate mitigation and assess the progress we have made in the effort to contain global warming to safe levels.

2. The Science

The danger of climate change arises because the build-up of relatively small proportions of gases in the atmosphere can restrict the emission of heat radiation from the earth to space, a process known as the greenhouse effect. The use of fossil fuels, coal, oil, and gas, causes about two thirds of the greenhouse gas emissions due to human activity. The most important gas is carbon dioxide, released during combustion, but escaping natural gas, or methane, is also significant. Other greenhouse gases include water vapour, nitrous oxide, and ozone, as well as various fluorinated gases. Practices in agriculture and forestry also affect the balance of greenhouse gas emissions, but energy is the main story, which is why energy policy is at the heart of climate mitigation.

Climate change is an enormously complex topic. There are many parts to the equation. They include the impact of the Sun's radiation on the earth and the uptake of heat by the atmosphere, the land, and the oceans. They include the tides and currents that distribute this heat. And there are the impacts of vegetation and of animal life. Some of the most powerful computers available are needed to run even simplified models of these interactions. So it's not surprising that there is a fair amount of uncertainty.

Nevertheless, some things are reasonably sure. A recent overview publication by the UK's Royal Society and the US National Academy of Sciences, the premier scientific bodies of the two countries, concludes that:

> The evidence is clear. It is now more certain than ever, based on many lines of evidence, that humans are changing the earth's climate. If emissions continue on their present trajectory without either technological or regulatory abatement, the warming of 2.6–4.8°C in addition to that which has already occurred will be expected by the end of the 21st Century. Global warming of just a few degrees will be associated with widespread changes in regional and local temperature and precipitation as well as with increases in some types of extreme weather events. These and other changes (such as sea level rises and storm surge) will have serious impacts on human societies and on the natural world.[i]

---

[i]Royal Society and US National Academy of Sciences, *Climate Change: Evidence and Causes*, February 2014.

A similar message comes from the Independent Panel on Climate Change of the UN. The IPCC is a scientific body with international membership that reviews and assesses the available scientific and technical information on climate change. Thousands of scientists worldwide contribute to its reports. In their latest summary report[ii] they say that global temperatures have already increased by 0.6–0.7°C between 1951 and 2010, largely due to greenhouse gas emissions. They conclude that if we continue with these emissions, on a "business as usual" basis, then global temperatures will have increased by 2°C by 2050, 3.7°C by 2100, 6.5°C by 2200, and 7.8°C by 2300. There is a considerable range of uncertainty around these estimates. For instance, the IPCC give a two thirds probability that the increase will be in the range of 2.6–4.8°C by 2100. Only if the world pursues "aggressive" mitigation strategies that would reduce emissions to near zero levels in 60 years can we contain the eventual temperature increase to 2°C.

In many ways, the increase in global temperature is a misleading measure of the impact of climate change. In the UK, for instance, the temperature frequently varies by more than 10°C from day to day. Is a 2–4°C change in global temperature such a big deal? It is the specific local consequences that count and, according to the IPCC, they are likely to be severe. For instance the sea level is set to rise by between 50 cm and 1 m by 2100 in "business as usual". Here are some of the key risks that they identify:

- Storm surges and flooding in low-lying coastal zones, towns and cities.
- Extreme weather events and extreme heat.
- Breakdown of food systems and rural livelihoods due to drought, especially in semi-arid regions.
- Loss of biodiversity, especially marine and coastal, particularly affecting fishing communities in the tropics and the Arctic.

[ii] IPCC, 2013: Summary for Policymakers. In: *Climate Change 2013*: *The Physical Science Basis*. Contribution of Working Group One to the Fifth Assessment Report of the Intergovernmental Panel on Climate Change, Stocker, T.F., D. Qin, G.-K. Plattner, M. Tignor, S.K. Allen, J. Boschung, A. Nauels, Y. Xia, V. Bex and P.M. Midgley (Eds.). Cambridge University Press, Cambridge, United Kingdom and New York, NY, USA.

Some of these risks are already judged to be considerable with a global temperature increase of 1–2°C, but they become "high to very high" with temperature increases of 4°C or more.[iii]

The IPCC say that many of the key risks will especially affect the least developed countries, given their limited ability to cope. The consequences for human societies are difficult to predict, but the report suggests with "medium confidence" that the consequent poverty and economic shocks may lead to displacement of people, violence, and civil war.

Once CO_2 has been released, it stays in the atmosphere for a long time. And changes in the earth's atmosphere also are very slow moving. This means that once the greenhouse gases have built-up to dangerous levels in the atmosphere the effects are irreversible, at least over a period of hundreds of years. According to the IPCC, if we are to limit global warming to 2°C, the target that global leaders have set, we can't afford to emit more than 2,900 billion tonnes of CO_2 in total. 1,900 billion tonnes of that has already been used up, since the industrial revolution, so we have 1,000 billion tonnes left. According to Carbon Brief,[iv] total CO_2 emissions in 2015 were about 40 billion tonnes (of which about 32 billion tonnes were from energy), and still on a gradually rising trend. As the IPCC have said, "the window for action is rapidly closing". We don't have the luxury of waiting to see how severe the effects of global warming are before taking action to reverse them. By then it will be too late.

Although the broad conclusion that climate change represents a grave risk is beyond reasonable doubt, there is still a lot of uncertainty about how and when the climate will change in different parts of the world and what the practical consequences of this will be. It is not always easy for scientists to deliver a message that is clear in its essentials but also recognises these uncertainties.

[iii] IPCC, 2014: Summary for policymakers. In: *Climate Change 2014*: *Impacts, Adaptation, and Vulnerability. Part A*: *Global and Sectoral Aspects*. Contribution of Working Group II to the Fifth Assessment Report of the Intergovernmental Panel on Climate Change [Field, C.B., V.R. Barros, D.J. Dokken, K.J. Mach, M.D. Mastrandrea, T.E. Bilir, M. Chatterjee, K.L. Ebi, Y.O. Estrada, R.C. Genova, B. Girma, E.S. Kissel, A.N. Levy, S. MacCracken, P.R. Mastrandrea, and L.L. White (eds.)]. Cambridge University Press, Cambridge, United Kingdom and New York, NY, USA, pp. 1–32.

[iv] Carbon Brief, *Carbon Countdown*, 19 May 2016.

3. Questioning the Scientific Consensus

Given the momentous implications of the scientific advice on climate change, and the fact that it threatens the future of some of the world's largest industries, it is not surprising that efforts have been made to discredit the work of the scientists involved. A number of contrarian scientists have been supported by conservative US think tanks and major energy companies.[v]

It is sometimes pointed out that the IPCC is "political". In one sense, how could it be otherwise? The IPCC was set up by the UN to advise on a matter which, without doubt, has major political implications. However, when one reads their report it is striking how careful they are to distinguish the strength of the evidence supporting their different conclusions as of high, medium, or low confidence.

Is it possible that some of the scientists working for the IPCC, or elsewhere, are making the most of the climate concerns because they are so anxious to get the attention of world leaders for what they regard as a critical issue? This was the issue raised when, in 2011, someone released more than 1,000 private emails of a leading climate research centre at the University of East Anglia in the run-up to a climate summit. Some of these suggested that scientists were reluctant to share certain data with people they viewed as wanting to make trouble and that they worked together to try to prevent papers offering different views of climate change being published. As one of the scientists acknowledged, some of the emails "did not read well".[vi]

An independent review, conducted in the light of these revelations, backed the integrity of the work of the scientists concerned, and did not find any evidence that might undermine the conclusions of IPCC assessments. But they did conclude that "there has been a consistent pattern of failing to display the proper degree of openness".[vii]

---

[v] Dunlap, R.E. *et al.*, *Climate Change Denial: Sources Actors and Strategies.* Routledge Handbook of Climate Change and Society, pp. 240–259, 2010.

[vi] Carrington, D. *What is Climategate? The Guardian*, Tuesday 22 November 2011.

[vii] *The Independent Climate Change E-mail Review*, Chair Sir Muir Russell, July 2010.

In the UK the most prominent figure arguing against action to mitigate climate change is Lord Lawson, a senior politician and former Chancellor of the Exchequer, or finance minister. He has attracted scientists and analysts who support his cause. But even he does not question the basic science of climate change, "The truth is that the amount of carbon dioxide in the world's atmosphere is indeed steadily increasing as a result of the burning of fossil fuels. ... And it is also a scientific fact that, other things being equal, this will make the place warmer".[viii]

The view of the scientific establishment is clear. Climate change is a grave threat. They are the best guide that we have. Is it possible that they are wrong? There are scientists who believe so. Even the IPCC's outlook includes, within the range of uncertainty, outcomes that are much more benign than their central case. It also includes outcomes that are much worse. The scientific establishment is not infallible and it should not be closed to debate with knowledgeable experts. Science proceeds through constant testing against evidence. But we would be crazy not to take climate change seriously on the grounds that our leading experts may be wrong or that we might get away with the most optimistic possibility in the range of uncertainty that they offer.

Some have pointed to an apparent pause in global temperature increases after 1998 as evidence that the climate scientists have got it wrong. However, the Royal Society and the US National Academy, in their report, conclude that:

> Since the very warm year 1998 that followed the strong 1997–98 El Nino, the increase in average surface temperature has slowed relative to the previous decade of rapid temperature increases. Despite the slower rate of warming the 2000s were warmer than the 1990s. A short-term slowdown in the warming of Earth's surface does not invalidate our understanding of long-term changes in global temperatures arising from human-induced changes in greenhouse gases.

[viii] Lord Lawson, *Daily Telegraph*, 28 September 2013. Available at http://www.telegraph.co.uk/earth/environment/climatechange/10340408/Climate-change-this-is-not-science-its-mumbo-jumbo.html. Accessed 17 June 2014.

Since then 2015 and 2016 have both turned out to be the warmest years on record, and this has taken some of the heat out of this debate. Indeed, with the exception of 1998, all 10 of the warmest years on record have been since 2000.

Even if we accept the existence, and the seriousness, of climate change, the question still remains, what should we do about it?

4. Responsibility to Act

Climate change raises particular ethical questions.[ix] Adults alive today are the last generation that has the opportunity of taking action to avoid the worst effects of climate change. If we decide to do so we will bear significant costs. The benefits, however, will be for future generations over several centuries, because most of the impact of climate change will be felt long after we are dead.

It is the rich developed nations that have caused a large part of the problem to date and it is these nations that have the capability and the resources to act. However, it is the poorer developing nations that are the most vulnerable to climate change both, in many cases, because of their geography, and because they are less capable of adaptation. As emphasised in the previous chapter, most of the future growth in carbon emissions is likely to come from major developing nations but their priority is immediate poverty relief and the raising of living standards, and not climate change.

Most ethical leaders, whether religious or secular, believe that we have an obligation to respond. For instance, in 2014 Pope Francis said, "On climate change there is a clear, definitive, and ineluctable ethical imperative to act".[x]

There are differing, although not necessarily contradictory, possible approaches to meeting our obligation to act on climate change. One approach is quasi-judicial, based on the "polluter pays" principle. On that basis, greenhouse gas emitting nations, and perhaps individual businesses and people, have an obligation to compensate others for the costs of those

---

[ix] Gardiner, S. *Ethics and Global Climate Change* Ethics **114**, April 2004, pp. 555–600.
[x] Pope Francis, *Message to the UN Convention on Climate Change*, 2014.

emissions to them. A particularly direct example of the case for compensation is the plight of small island states threatened with being submerged.

Others take a quasi-religious approach in which it is the sanctity of the earth, or, put another way, our stewardship for the future of the earth, that is paramount. They tend to put more emphasis on protecting the natural world and the diversity of species, not just human welfare in its narrow sense. On that approach we have an absolute duty to minimise the impact of climate change.

A third approach is somewhat more flexible. This is that we have a duty to hand on to future generations a total inheritance that is at least as favourable as that which we received. This inheritance certainly includes the state of the environment but it also includes human and physical capital including education, infrastructure, and technology.

In 2006, the UK government commissioned its own global study of the case for action on climate change. This was the Economics of Climate Change, generally known as the Stern review, after its author Lord Nicolas Stern. Stern attempted to quantify both the avoidable costs of global warming under "business as usual" and the costs of mitigation. As he acknowledged, this required some "heroic" assumptions. He came to the conclusion that, if we do nothing, the costs of climate change will be equivalent to losing at least 5% of the value of the global economy each year and possibly as much as 20%. In contrast, the costs of action to avoid the worst impacts can be limited to around 1% of the GDP each year. The risks are, "on a scale similar to those associated with the great wars and economic depression of the first half of the 20th century". If we do not take action in the next 10–20 years "it will be difficult or impossible to reverse these changes". It is already 10 years since this report was published so, on this basis, time is definitely not on our side.

As you would expect, with such a salient study, Lord Stern's work has been widely critiqued.

The strongest area of criticism has been on the rate of interest, or "discount" rate that he applies to future benefits. This may seem rather technical but in practice, if one follows the welfare economics approach adopted by Lord Stern, it is at the heart of the issue. This is because almost all the economic losses will occur after 2100 and most of them after 2200, whereas action to mitigate climate change needs to start now.

Lord Stern used a discount rate of 1.4% p.a. To arrive at this, he starts from the principle that a future generation has the same claim on our ethical consideration as the current one. So there is no discounting simply due to the passage of time. But he allows two reasons for valuing benefits in the distant future less than those today, on principles known to economists as the "Ramsey Formula". The first is that future generations may not exist if some cataclysmic event takes place. The second is that if future generations are much richer than we are today then marginal increases or reductions in living standards may be less important to them than they would be to us. Lord Stern's conclusion is not very different from the British government's rules for evaluating government projects, which recommend discount rates of 3.5% for projects with benefits in the next 30 years but only 1.5% for benefits arising more than 200 years from today.

One of the more prominent critics of Lord Stern was William Nordhaus.[xi] In an academic paper of 2007 he begins by questioning Stern's overall "prescriptive" method. "The Review takes the lofty vantage point of the world social planner, perhaps stoking the dying embers of the British Empire". Nordhaus uses a smaller model of the economics of climate change which he adjusts to parallel the assumptions in the Stern report and he runs it with a discount rate that he considers to be consistent with market interest rates. The implied costs of carbon and the optimal rate of emissions reduction are well below those of the Stern Review. Nevertheless, "While the findings of such mainstream economic assessments may not satisfy the most ardent environmentalists, they will go far beyond the meagre policies currently in place".

It is easy to see why the discount rate is so important in this kind of assessment.

At a discount rate of 3% p.a. the present value of £1,000 one hundred years from now is £52. At 8% p.a. it is 45 p. Thus if one applies a discount rate in the range required for commercial projects (6–7% p.a. upwards, depending on the risk) the present value of any benefit to be obtained more than 50 years into the future is practically zero.

[xi] Nordhaus, W. (2007). *A Review of the Stern Review on the Economics of Climate Change*, *Journal of Economic Literature* B **42** (September 2007), pp. 686–702.

The argument for applying commercial style interest rates is that, if the rate of benefits from climate mitigation is lower, then future generations would be better off if the same amount of money was spent on commercial investment rather than protecting the environment. However, this depends on the assumption that investment in the environment will automatically reduce the pool of money available for commercial investment, which is not necessarily the case.

As Martin Weitzman has observed[xii]:

> The most critical single problem with discounting future benefits and costs is that no consensus now exists, or for that matter ever has existed, about what actual rate of interest to use. Economic opinion is divided on a number of fundamental aspects ... All these, and many more, considerations are fundamentally matters of judgement or opinion, on which fully informed and fully rational individuals might expect to differ.

The best arguments to be made against action to address climate change are these. Firstly that there are other forms of investment for the future, for instance in education or infrastructure, which will ultimately benefit future generations more. Secondly that, assuming continuing economic and technical progress, future generations will be richer than we are and more capable of addressing the problems of climate change.

These arguments would be more compelling if we believed that future generations, at a cost, would have the opportunity to reverse climate change. Unfortunately, that is not the case. Climate change, once it has taken place, will be irreversible over a long period. This means that, to make these arguments, we would have to be willing, on behalf of future generations, to trade environmental degradation for other benefits. Many people in the West would not be willing, in principle, to make such a trade, although the case might appear different to people living in poverty in the developing world. However, the argument also depends on the assumption that climate change will not be so cataclysmic as to undermine the economic prospects of future generations, and that is highly questionable.

[xii]Weitzman, M. (2001). Gamma Discounting. *American Economic Review* **91**, 1 March 2001, pp. 260–271.

The fundamental question on climate change is this. Are we willing to hand on to future generations a climate that has been altered, through human intervention, in ways that appear certain to be seriously harmful and may be catastrophic? It comes down to a visceral judgement. Most people will say no, and that is certainly my conclusion.

5. Counting the Cost

The next question is, "how much should we be prepared to pay" or, closely related, "how much effort are we willing to make"? I have mentioned that Lord Stern, in his original study, estimated the cost of addressing climate change at around 1% of world GDP. Since then, Lord Stern has raised this estimate to 2%.[xiii] The Grantham Institute at Imperial College has estimated the cost at 1.7% if we start in earnest from 2020 but 2.2% if we delay until 2030.[xiv]

The 2014 Synthesis Report of the IPCC recognises that estimates of the economic costs of mitigation vary widely depending on methodologies and assumptions. They take scenarios in which all countries begin mitigation immediately, in which there is a single global price of carbon, and in which all technologies are available, as their cost effective benchmark. This is obviously an optimistic, not to say unrealistic, scenario. In that case they estimate the total loss of global consumption at 1–4% in 2030, 2–6% in 2050, and 3–11% in 2100.[xv]

However, simply adding up the extra costs of low-carbon energy is a rather simplistic way of looking at a major transition with many implications. Lord Stern has recently argued, based on a paper by his team at the Grantham Institute,[xvi] that the total impact on the economy is much more positive. This paper attempts to assess the co-benefits of climate change

---

[xiii] Cost of tackling climate change has doubled, warns Stern. *The Guardian*, 26 June 2008.
[xiv] Avoid 2 (partnership of the Met Office, the Grantham Institute Imperial College, the Tyndall Centre, and the Walker Institute), *Can We Avoid Dangerous Climate Change?* 2017.
[xv] IPCC, *Synthesis Report*, 2014.
[xvi] Green F. *Nationally Self Interested Climate Change Mitigation: A Unified Conceptual Framework.* Grantham Research Institute on Climate Change and the Environment, July 2015.

measures. These include reductions in indoor and outdoor air pollution, and many efficiency improvements. The elimination of subsidies for fossil fuels also improves economic performance. And support for advanced low-carbon technologies has spin-offs for innovation more generally and reducing costs for the future. Taking all these factors into account the paper argues that there is a net national benefit from carbon reduction measures, even before the mitigation of climate change is taken into account.

In their 2014 publication, Energy Technology Perspectives, the IEA compared the total additional investment required for their 2°C scenario with the ultimate benefits in terms of reduced fuel costs. The balance turned out to be positive, even at a discount rate of 10% p.a. This does not necessarily mean that there was a positive return on low-carbon technology because the scenario also included big improvements in energy efficiency. Also, savings in fuel costs to consuming nations are not necessarily a benefit to the world as a whole because a large part of the cost of fossil fuels, especially oil, goes in rent to the lowest cost producers. Nevertheless, all these studies strongly reinforce the logic of pursuing a low-carbon transition.

The idea that climate mitigation can make a positive contribution to the economy has been captured in the term "green growth". This was first adopted at a meeting of Asia Pacific leaders in 2005, but was subsequently adopted by high level meetings of the UN and by the OECD. It reached its apogee following the financial crash of 2008, when investment in green energy was advocated as a means of lifting the global economy.

Climate change is an inspiring cause, as anyone who works with students in a Western university will be aware. The economy is not a zero sum game and is ultimately driven by what Maynard Keynes memorably described as "animal spirits". The drive for a less polluted world is definitely one of the forces capable of renewing and reinvigorating modern economies.

Many will say that just 1% or 2% of the economy is, anyway, a small price to pay for saving the planet. Certainly that was the view of politicians when costs of this order were first presented by the IEA. However, the costs of policies are not usually presented in terms of percentages of GDP. Even 1% of world GDP is a lot of money, about $700 billion in 2014. It seems an even larger sum when you ask the question, who is going to pay?

The developing countries, where much of the expenditure is needed have made it clear that they expect support from the developed world. However, the developed world is having difficulty in providing even the $100 billion p.a. from 2020, that was first promised at the 2009 climate summit in Copenhagen. Most of that is now expected to be commercial or semi-commercial finance. Even within developed countries, in Prime Minister May's memorable expression, there is a large population that can "just about manage".[xvii] There is limit to what they will be willing or able to pay. If you accept that those suffering fuel poverty must be protected, as well as energy-intensive industries facing international competition, the pool of those asked to bear the cost is looking rather narrow.

Viewed from the perspective of government policymakers, many of the measures that are effective in reducing climate emissions, whatever their longer term benefits, have the immediate effect of imposing additional costs on energy consumers or taxpayers or of reducing tax revenues. This is true of subsidies for renewables, nuclear power or electric vehicles, or incentives for energy efficiency. It means that climate measures are competing for funds with other claims on public finances such as education, health, infrastructure, and deficit reduction. In some of these areas also it can be claimed that increased spending today could have disproportionate benefits for the longer term and improve the quality of life. Finance officials inevitably become somewhat hardened to these arguments. Most governments recognise the importance and positive value of achieving a low-carbon energy transition but finding the resources to act on the scale required is more difficult. There is no disguising the fact that tough decisions are required.

6. A Simpler Life

There are some who believe that the West has already become far too materialistic for our own good. We would be happier and more fulfilled if we could escape the treadmill of working ever longer hours to afford "trophy" possessions and instead put more of our efforts into community, social, and

---

[xvii]Theresa May, First Statement as Prime Minister, delivered in Downing Street, 13 July 2016.

leisure pursuits that, amongst other benefits, do not make such demands on natural resources. This is the argument that the poet Wordsworth put in his sonnet, written at the time of the first industrial revolution:

> The world is too much with us. Late and soon, Getting and spending, we lay waste our powers[xviii]

Prosperity Without Growth is a highly articulate and popular book which presents this argument.[xix] From this perspective the need to reduce carbon emissions is not a burden on economic growth, but one of many positive consequences of a fundamental change in lifestyle that is very much in our own interests.

Most people will agree with the central message of *Prosperity Without Growth*. We should lead less materialistic and more fulfilling lives. To a limited degree perhaps, this is already happening in the most developed countries. The current debate on "work-life balance" is obviously related. The service sector now dominates the economies of the rich OECD countries and, since it is much less energy intensive than manufacturing and construction, this is one of the reasons why the carbon emissions of these countries are now more or less flat and are set to decline. Energy efficiency, which in its widest sense can be taken to include a switch to a less energy-intensive economy, is recognised to be the most important single option for carbon reduction.

But as a solution to the challenge of climate change, this sort of approach has its limitations. As religious and ethical leaders have found down the ages, it's not easy to change people's lifestyles. Certainly it's perilous for the state to get involved, witness the cries of "nanny state" that arises when the UK government takes even the most modest steps in this direction. As a result of efforts to persuade people to smoke and drink less, and to eat healthier food, Britain came third from top in the Institute of Economic Affairs' 2016 "Nanny State Index".[xx]

[xviii] Wordsworth Sonnet. *The World is Too Much with us Late and Soon.*

[xix] Jackson, T. *Prosperity Without Growth*, Earthscan/Routledge, 2009.

[xx] The Institute of Economic Affairs, *Britain Third Worst Country in EU for Nanny State Regulation*, 31 March 2016.

But there is a more fundamental reason why we cannot rely on changes in lifestyle to curb world emissions. Most of the world's CO_2 emissions, and all the projected future growth in emissions, are not in the rich, "materialistic" developed world, but in the developing nations.[xxi] *Prosperity Without Growth* points out that for nations that have achieved middle income status (£15,000 per person) there doesn't seem to be much correlation between further increases in income and human welfare measures such as life expectancy and stated contentment. But this is definitely not the case for poorer countries, for obvious reasons. These countries are still building their basic capabilities for health services, sanitation, transport, food distribution, education, electric power etc. All this is energy intensive. We cannot rely on lifestyle changes to solve the core problem of energy for economic development.

7. Global Cooperation

The international climate treaty, the United Nations Framework Convention on Climate Change (UNFCCC), adopted in 1992, is the foundation of international climate negotiations. It is an uplifting document.[xxii] It recognises the danger of climate change and sets the objective of "stabilization of greenhouse gas concentrations in the atmosphere at a level that would prevent dangerous anthropogenic interference with the climate system". Nations have "common but differentiated responsibilities" for achieving this. The developed nations must lead in taking action and must also give financial help to the developing nations to enable them to follow. Action by developing nations depends on the receipt of this financial help. For developing nations, as previously noted, "economic and social development and poverty eradication are the first and overriding priorities".

Some 200 nations, including all major developing and developed nations, have ratified the UNFCCC and this is undoubtedly a major achievement of international diplomacy. However, as the title says, this is a framework document only. It does not contain any specific national targets or obligations.

[xxi] International Energy Agency, *World Energy Outlook*, 2013.
[xxii] United Nations, *Framework Convention on Climate Change*, New York, 1992.

The first, and still the most important, effort to attach specific national obligations to the UNFCCC was the Kyoto Protocol[xxiii] of 1997. The Protocol contained targets for each of the developed nations to reduce their average greenhouse gas emissions in the period 2008–2012, compared to 1990. The structure was flexible and included trading mechanisms. Any of the developed nations could agree together to meet their targets jointly and to trade emissions reductions from specific projects in a system called "Joint Implementation". They could also purchase emissions reductions from projects in developing countries, with no quotas of their own, under the "Clean Development Mechanism" (CDM).

To qualify under the CDM scheme emissions savings have to be authorised as Certified Emissions Reductions (CER), a process that is overseen by an international CDM Executive Board. The mechanism has been much criticised, as one would expect of an administered scheme of this kind. Are the projects genuinely additional? Are the carbon reductions in the most worthwhile areas? Is there too much bureaucracy or, conversely, not enough control? These are all difficult issues to manage. However, according to the Executive Board the scheme has achieved 1.38 billion tonnes of certified CO_2 equivalent emissions and facilitated $315 billion of capital investment, mainly in China and India.[xxiv]

Compliance with the Kyoto targets has been patchy.[xxv] For Russia and Ukraine, compliance was easy as a consequence of the collapse of the Soviet Union and a sharp decline in heavy industry. The EU as a whole has met its target, helped by the decline of the coal industry and the "dash for gas" in the UK and elsewhere, by the collapse of the heavy industrial sector in East Germany, and also by aggressive programmes for renewables and energy efficiency.

The US never ratified the Protocol, although US officials had played a major part in its negotiation, and therefore was never committed to its

---

[xxiii] United Nations, *Kyoto Protocol to the United Nations Convention on Climate Change*, 1998.

[xxiv] Annual Report of the Clean Development Mechanism 2013, UN Framework Convention on Climate Change.

[xxv] Haita, C. The State of Compliance in the Kyoto Protocol, International Center for Climate Governance, 2013.

national target. Until recently, it appeared that US emissions would far exceed it. But more recently the displacement of coal by shale gas in US power stations has sharply reduced US emissions and it now seems likely that the US may, fortuitously, have met its target. Australia also met its Kyoto target, although this admitted some increase in emissions levels. Other countries have not fared so well. Canada's emissions far exceeded its target and Canada formally pulled out of Kyoto in 2011.[xxvi] Japan's actual emissions rose, but Japan nevertheless met its Kyoto target through a major and costly programme of purchasing international credits.[xxvii]

With the expiry of the first Kyoto commitment period in 2012, developing countries felt strongly that a continuation was essential to demonstrate the leading role of developed nations under the UNFCCC. Such a continuation was indeed negotiated with new targets for the period 2013–2020 but far fewer countries signed up.[xxviii] Essentially it is now mainly the EU that accepts specific Kyoto targets for its emissions. This is a disappointing outcome, considering that the EU now accounts for only about 12% of global carbon dioxide emissions.[xxix]

Since the UNFCCC was first adopted, annual climate summits, or "Conferences of the Parties (COP)" as they are more formally known, have tried to build the international machinery for achieving its objective of mitigating climate change.

8. Copenhagen Climate Summit

The Copenhagen COP of 2009 was particularly significant. Expectations were high in the run-up to the meeting, as it became clear that the leaders of all major nations, including President Obama and Chinese Premier Wen

---

[xxvi] BBC News, US and Canada, 13 Dec 2011. Available at http://www.bbc.co.uk/news/world-us-canada-16151310. Accessed 17 June 2014.

[xxvii] Reuters US Edition, 17 November 2013, Japan Uses offsets to meet Kyoto emissions goal. Available at http://www.reuters.com/article/2013/11/17/us-climate-japan-co-idUSBRE9AG02420131117. Accessed 17 June 2014.

[xxviii] Carbon Trust, 23 Jan 2013 Doha: It kept the show on the road but only just. Available at http://www.carbontrust.com/news/2013/01/doha-it-kept-the-show-on-the-road-but-only-just. Accessed 17 June 2014.

[xxix] International Energy Agency, *World Energy Outlook*, 2013.

Jiabao would attend. This was supposed to be the moment when tough decisions would be made on the implementation of the UNFCCC, to be followed shortly by legal agreements.[xxx] In fact it was a disaster, at least in terms of public relations.

Usually for international meetings at this level the main conclusions are prepared by staff of the organisation, in consultation with top-level officials of national governments, in advance. Leaders can add final touches in their discussions and their personal chemistry. But they are not expected to do the heavy lifting around the table, and certainly not at an event such as Copenhagen, attended by 115 nations. That is why the senior officials who prepare for G20 summit meetings are known as "sherpas", a reference to the Nepalese mountaineers who support expeditions to Mount Everest. The joke has perhaps worn thin over the years. But the point is valid. Unfortunately, the Copenhagen summit was not well prepared in this way.

The talks did not go well. Developing nations wanted the developed nations to commit to a tough extension of the Kyoto protocol or similar legally binding cuts and they wanted specific large scale commitments of financial help. The developed nations wanted the developing nations to start sharing the load. The meeting was in chaos and heading for complete failure. According to US news reports,[xxxi] about an hour after he was originally supposed to leave, President Obama learned that the leaders of China, India, South Africa and Brazil were meeting. He gatecrashed the event and in 45 min of intensive discussion they came up with a non-binding political agreement, now known as the Copenhagen Accord.[xxxii]

The Accord contained a strongly worded statement of political will to combat climate change. It entrenched a specific target of limiting the global temperature increase to 2°C. And it committed the developed

---

[xxx] Jake Schmidt, *Natural Resources Defence Council Staff Blog*, 30 November 2009. Available at http://switchboard.nrdc.org/blogs/jschmidt/copenhagen_part1.html. Accessed 24 June 2014.

[xxxi] Lindsey Ellerston, *ABC News*, 18 December 2009, High Drama in Copenhagen. Available at http://abcnews.go.com/blogs/politics/2009/12/high-drama-in-copenhagen-per-administration-officials/. Accessed 24 June 2014.

[xxxii] United Nations, FCCC/CP/2009/11 Add.1 30 March 2010. pp. 4–9.

countries to "mobilizing jointly" $100 billion p.a. by 2020 to address the needs of developing countries. But there were no binding national emissions targets and no collective process to put them in place. To some extent, honour had been saved. But the verdict of the chief negotiator of the group of developing nations was that the Accord was, "The lowest level of ambition you can imagine".[xxxiii] The Executive Director of Greenpeace UK was more outspoken, "The city of Copenhagen is a crime scene tonight, with the guilty men and women fleeing to the airport".[xxxiv]

The management of expectations at Copenhagen was a disaster. Because expectations were so high, and because the psychology had seemed to be building towards a really big step forward, a relatively modest outcome was widely perceived as an almost complete failure. Certainly momentum was lost. As the Wall Street Journal reported, the businesses who would need to invest to create a lower carbon future had hoped for a clear signal but were instead facing continued uncertainty. "CO_2 Pact Leaves Businesses Up in the Air" was the Journal's headline.[xxxv]

But in the cold light of day we can see that Copenhagen did make some important positive contributions and may have ushered in an era of greater realism. The most important of these contributions came, in fact, before the meeting even started. Almost all the major players offered their own voluntary national targets for managing their carbon emissions. For instance, China undertook to reduce carbon emissions per unit of GDP by 40–45% in 2020 compared to 2005, and India made a similar commitment to reduce emissions intensity by 20–25%. Some were inclined to dismiss these offers, on the grounds that they contained no absolute limit on

[xxxiii] The Guardian, *Low Targets, Goals Dropped: Copenhagen ends in Failure,* 19 December 2009.

[xxxiv] John Vidal *et al.*, *The Guardian*, 19 December 2009. Available at http://www.the guardian.com/environment/2009/dec/18/copenhagen-deal. Accessed 24 June 2014. http://online.wsj.com/news/articles/SB126118612845198057?mg=reno64-wsj&url=http%3A%2F%2Fonline.

[xxxv] Chazan, G., *Wall Street Journal*, 21 December 2009. Available at http://online.wsj.com/news/articles/SB126118612845198057?mg=reno64-wsj&url=http%3A%2F%2Fonline. wsj.com%2Farticle%2FSB126118612845198057.html. Accessed 25 June 2014.

emissions. However, studies at the Grantham Institute[xxxvi] have shown that they were set at about the level one might expect, so as to represent a significant improvement on business as usual, but to be reachable at manageable cost through vigorous policy measures.

In the analysis that they released in 2010[xxxvii] the International Energy Agency tried to estimate the impact of the national commitments that were made at Copenhagen and how they measured up to the global 2°C target. They concluded, broadly, that without these pledges the world was heading for a temperature rise in excess of 6°C but that if the pledges are met, the outlook is for a temperature increase of above 3.5°C. This, of course, is not enough. But the Copenhagen pledges represented a major step forward.

9. Paris Climate Summit

The next big step after Copenhagen was the Paris COP of December 2015. The event was foreshadowed by an agreement at the Durban COP in 2011 that the parties would agree an "outcome with legal force" no later than 2015.

The lessons of Copenhagen seem to have been learned at Paris. An extensive and inclusive programme of drafting the conclusions was instigated well in advance. As the COP approached, the broad outline of what was to be expected was already clear.

The Paris event also benefited from a following wind in the form of an impressive commitment of the two most important players, Presidents Obama and Xi Jinping. Several factors had put President Obama on the front foot on carbon reductions. These included tighter US vehicle standards and the displacement of coal by shale gas in US power stations. They also included the Clean Power Plan for the reduction of carbon dioxide emissions from power stations being pursued by the US's environmental regulator, the Environmental Protection Agency (EPA), with the President's strong support. From Xi Jinping's perspective, besides China's own

---

[xxxvi] Hirst, N. *et al.* 2012. *An Assessment of India's 2020 Carbon Intensity Target*, Grantham Institute for Climate Change Report GR4 and *An Assessment of China's 2020 Carbon Intensity Target*, Grantham Institute for Climate Change Report GR1.
[xxxvii] International Energy Agency, *World Energy Outlook*, 2010.

vulnerability to global warming, a positive stance on climate change fitted in well with China's ambition to rebalance its economy towards less energy intensive and polluting industries.

In November 2014, during a state visit of Xi Jinping to the US, the two presidents made a joint statement in which they described climate change as "one of the greatest threats facing humanity" and committed themselves to work for an ambitious summit outcome.[xxxviii]

President Obama announced a target of reducing US emissions by 25–28% between 2005 and 2025. President Xi Jinping set targets that Chinese carbon dioxide emissions will peak around 2030, with best efforts to peak early, and that China also intends to increase the share of non-fossil fuels in energy consumption to around 20% by 2030. Bearing in mind that the European Union was already strongly committed to the success of Paris, this was a major boost for climate negotiations.

At the summit, nearly 200 nations, including all major emitters, signed a legal agreement. The aim of the agreement is to hold the increase in global average temperatures to well below 2°C above pre-industrial levels and to pursue efforts to limit the temperature increase to 1.5°C.

The most important advances made at the summit were the agreement that all parties, and not just the developed countries, should aim for their emissions to reach a peak as soon as possible, that they must submit their "intended nationally determined contributions" (INDCs) to the global response to climate change, and that these must be renewed every 5 years with the aim of progressively ratcheting up. The agreement includes provisions to improve the quality, comparability, and transparency of these contributions.

Since then nearly 200 countries have submitted their INDCs. In the case of most developed countries these provide for absolute percentage reductions in emissions. For instance, the EU has said that it will reduce greenhouse gas emissions by at least 40% by 2030 compared to 1990. For most developing countries they provide either for a reduction per unit of GDP or for a reduction compared to an estimate of business as usual. For instance India has said that it will reduce the emissions intensity of its GDP by 33–35% by 2030 compared to 2005.

[xxxviii] The White House, *US–China Joint Announcement on Climate Change*, 12 November 2014.

The agreement still recognises that it is for the developed countries to take the lead, reflecting the principle of "common but differentiated responsibilities" in the original convention. The developed countries are required to set absolute economy-wide emission reduction targets, whereas the developing countries are given more flexibility on the nature of their contributions for the time being.

The agreement recognises the need for the developed countries to support the developing countries. However, the figure of $100 billion p.a. does not appear in the legally binding section. The agreement also recognises the loss or damage that may result from climate change but says, explicitly, that these provisions do not provide the basis for any liability or compensation.

The Paris climate summit was undoubtedly an important success for climate negotiations. It was well attended by the heads of government of leading nations and some 200 nations committed themselves to emissions reduction targets and to positive agreed conclusions on the way forward. However, this success was achieved at the price of accepting more modest, but perhaps also more realistic, ambitions as to what could be achieved through the international process.

Contrary to the spirit of Durban, there were no legally binding commitments to reduce emissions. Paris has clearly established that the approach will be "bottom up" rather than "top down". The attempt that was made at Kyoto to impose centrally determined and legally binding quotas has been abandoned. Instead, as the expression "intended nationally determined contributions" implies, each nation will set its own non-binding targets. It seems clear that, at least for the time being, many of the most important governments are not willing to submit the level of their carbon emissions, and thus a large part of their energy policies, to international control.

The INDCs, taken together, are not sufficient to achieve the stated objective of limiting global warming to 2°C, much less to 1.5°C. This is recognised explicitly in the Paris conclusions, which call for ratcheting up in the future.

The Paris agreement was well received by the media and, in contrast to Copenhagen, has imparted a positive momentum to the climate process. But there is also a risk, because of this positive reception, of giving the impression that the problem has been solved. In reality, there is a long and rocky road still to be navigated.

The tensions in climate negotiations will remain. The developing nations expect the rich developed nations to take the lead, not only in setting legally binding targets for emissions reduction but also in paying for climate measures in poorer countries. The developed nations have set specific emissions reduction targets but for the most part these are not legally binding. Although they accept in principle the need to help the poorer countries, they have avoided specific commitments. The figure of $100 billion which was first mentioned in the Copenhagen Accord is not very meaningful because the developed nations have avoided specifying how much of this will be government grants and how much will be from a "wide variety of sources instruments and channels", which could include commercial loans. Nor is there any commitment as to how much of this money will go through the mechanisms of the UNFCCC Treaty, which would enable the developing nations to share in its management.

The developed nations want the developing nations to move as soon as possible to setting their own targets for the absolute emissions reductions, bearing in mind that they now account for the majority of global emissions and all the growth in emissions. Developing nations are determined not to inhibit their growth potential and, therefore, have generally set targets that are relative either to GDP or to projections of business as usual. Many of the national targets set by developing countries are also conditional on receiving sufficient financial support.

Small island states, most directly threatened by rising sea levels, are pressing for tougher global warming targets. It is largely because of pressure from them that "efforts to limit the temperature increase to 1.5°C" are promised in the Paris agreement. At the moment 1.5°C seems far out of reach. A recent statement by the leader of their association, Thoriq Ibrahim dramatised their situation. "the implementation of the Paris Agreement ... is the last hope for the survival of our members ... since we last met in Morocco, the stakes for our group were tragically underscored by the loss of five islands in the Solomon Islands archipelago due to sea level rise".[xxxix]

[xxxix] Press release of the Alliance of Small Island States (AOSIS), *Small Islands Highlight Urgency and Benefits of Climate Action in Bonn*, 1 May 2017.

Without getting too starry eyed we should also recognise the UNFCCC negotiations, difficult as they are, as a step along the road of human progress. For the first time virtually all nations, rich and poor alike, are locked in serious negotiation about the future of the planet that we share. That is surely a significant landmark.

10. US Withdrawal from the Paris Agreement

President Trump's announcement, in June 2017, that the US will withdraw from the Paris agreement must be regarded as a major setback. It means that US leadership will be absent, at least during Donald Trump's Presidency, from the vital process of building the agreement and ratcheting up national commitments. It implies that the US may not now meet its own voluntary emissions target. It also calls in question, as already mentioned, the promised $100 billion p.a. of support to developing countries.

However, it is encouraging that other participants, notably China and the EU, have reasserted their commitment to the agreement, following President Trump's announcement. There seems no doubt that it will continue in the absence of the US, and will still be a powerful influence for emissions reduction.

In his statement on the Paris agreement, President Trump did not express scepticism on climate science, though he has done so in the past. Nevertheless, he is already, in June 2017, taking actions that will impinge on the US climate effort. These will impact primarily on federal environmental regulation and spending. According to the *New York Times*, in appointing Scott Pruitt to run the EPA, President Trump is "putting a seasoned legal opponent of the agency at the helm of [his] efforts to dismantle major regulations on climate change and clean water — and to cut the size and authority of the government's environmental enforcer".[xl] Key environmental regulations are being rescinded. This includes President Obama's Clean Power Plan, which would have set limits on power sector emissions state by state. Major budget cuts are planned for the EPA. All this is going to have serious environmental consequences.

[xl] New York Times, *Senate Confirms Scott Pruitt as EPA Head*, 17 February 2017.

However, President Trump has presented himself as a friend to the renewables as well as to fossil fuels. Many of the incentives for renewables and clean energy in the US are at state, rather than federal, level, and economics are moving in favour of renewables. So the renewables industries in the US will no doubt continue to thrive.

The main reassurance, in all this negative news about climate mitigation in the US, is that the costs of wind and solar energy continue to decline and that their future in the US seems bright, almost irrespective of federal government policy.

11. Energy Governance

The failure of the "top down" approach to climate negotiations means that progress now depends on the voluntary offers of individual nations to reduce their carbon emissions. Since two thirds of emissions are energy related this will largely depend on their energy policies, and how successfully they can combine carbon reduction with other objectives such as energy security, affordability and economic growth and development. Peer pressure at the UNFCCC will continue to be important. But we also need to intensify international debate on the specific policies and technologies that can be most effective. Unfortunately, the main body for this kind of engagement, the IEA, still excludes developing nations from full membership. Chapter 10 discusses the current state of global energy governance and the reforms that are urgently needed to make it fit for purposes to tackle today's energy challenges.

12. The Current State of Climate Mitigation

How much progress have we really made towards reducing greenhouse gas emissions from energy?

Since the signing of the climate treaty in 1992, at which world leaders agreed to stabilise greenhouse gas concentrations at safe levels, world carbon emissions from energy have increased by more than 50%. They stabilised in 2015 and 2016 but it is far from clear that they have yet turned the corner. The developed (OECD) countries account for a bit less than one third of emissions. The developing countries plus Middle East oil producers account for the other two thirds. Emissions from OECD countries seem set for steady decline. But emissions from developing countries continue to grow.

The INDCs, if delivered, will have a strong impact. They will reduce the rate of increase in emissions to one third of that experienced since 2000. And they will substantially decouple emissions growth from economic growth. In 2030, emissions per unit of economic output will be 40% lower than today. However, to meet the 2°C target, global emissions would need to peak before 2020 and then decline steeply to very low levels by 2050, whereas on a fairly optimistic view of the impact of the INDCs, they will still be increasing in 2030, albeit rather slowly. Energy-related carbon dioxide emissions are projected at 35 billion tonnes in 2030, whereas for the IEA's 2°C case, they need to be at 19 billion tonnes. The IEA project that the resultant increase in global temperature would be 2.7°C. So there is still a big gap.

To meet the climate challenge we need to improve energy efficiency and switch to less carbon-intensive energy sources at a rate that more than compensates, in terms of emissions, for the impact of continuing economic growth, especially in the developing world. Everyone agrees that this is technically possible, but that an ambitious transformation is required. The biggest part is to improve the energy efficiency of all the main areas of energy use, power generation, heating of buildings, transport, and industry. This is discussed in Chapter 8 on energy efficiency. But switching to lower carbon sources of energy is almost as important, especially to drive down emissions from coal, which is the biggest polluter. To some extent this can be done through more efficient coal stations or switching to gas, but the longer term solutions are renewables, nuclear power, or emissions storage.

Renewables will have an important role and their costs are continuing to fall. But today their impact is still fairly modest. The record level of new wind and solar capacity installed in 2015 represented about 2% of global generating capacity, about the same as the annual rate of capacity growth. This is having a significant impact on power generation, but so far low-carbon technologies have not really penetrated building heating, transport, or industry, sectors which make up the majority of energy consumption, to any significant degree, and this is the next big challenge. The potential for low-carbon technologies, especially renewables, to achieve a revolution in these areas is discussed in the next chapter.

13. Developing Nations

Although the developing nations account for most of today's emissions, and all the expected growth in emissions, it is important to remember that most of the greenhouse gases that are in the atmosphere today were put there, historically, by the developed nations. Also, most importantly, each person in the developed world still emits, on average, more than twice as much CO_2 as people in the developing world. For instance India's emissions per head, although rising, are currently less than 15% of the average for the OECD. It is reasonable enough that the developing world expects the developed nations to give a lead and take the first action to reduce their high levels of emissions. However, the raw statistics show that early action by developing nations will also be essential if we are to avoid the grave consequences of climate change.

The critical question, therefore, is how developing nations can moderate and eventually reduce their greenhouse gas emissions in a way that positively supports their ambitions for economic growth, poverty eradication, and rising living standards. Greater realism is needed in addressing this issue. As I pointed out above with the example of South Africa, where developing nations face a critical and urgent need for a rapid increase in energy supply we cannot expect all of this to come from low-carbon sources and energy efficiency improvements. Developed nations themselves are struggling to make this transformation and are finding that the costs are high and that ambitious efficiency targets are difficult to achieve. Gas and even, in some circumstances coal, may still be needed as part of plans for growth and social justice in developing nations. The role of developed nations and of international financial institutions should be to support these developments, to ensure that new fossil plants, where they are necessary, are as efficient as possible (and where possible are capable, later on, of having emissions captured for underground sequestration), and that this fossil plant is part of longer term plans for greater efficiency and low-carbon transformation of the energy sector.

In 2017 a row erupted between the World Bank Information Centre, which monitors the World Bank on behalf of NGOs, and the Bank itself. The Centre complained that the Bank was "introducing new fossil fuel subsidies, undermining its own climate change commitments and forest protection efforts". The claim was that, in its Development Policy Finance programme for helping poorer nations to progress, the Bank had sometimes given its support to development programmes that included an element of fossil fuels. For instance it is claimed that programmes supported by the World Bank include support for coal or gas developments in Peru, Indonesia, Egypt, and Mozambique.

In its reply the World Bank does not deny that fossil energy sometimes benefits from these programmes, thought it says that the Information Centre report is inaccurate. "Helping countries make the transition to clean energy is core to our work, in addition to securing affordable, sustainable and reliable energy services to the more than 1 billion people who currently lack access". In Egypt the Bank's engagement is focused on "eliminating extreme poverty and boosting shared prosperity in a sustainable manner". Mozambique is one of the world's poorest countries, but is rich in natural gas. 25% of the population have access to electricity. In remote off-grid areas the World Bank is promoting solar energy. But in both Egypt and Mozambique the Bank is supporting the overall efficiency of the energy systems, including gas development.

The World Bank has the better of this argument. They should definitely be promoting the low-carbon transition, as they are clearly doing. But in countries with extreme poverty weight also has to be given to finding the most immediate and practical means to raise living standards, and that may involve some element of fossil energy. The Bank also has to work with the governments of the countries concerned and their priorities must be given weight.

14. Stranded Fossil Fuel Reserves

In recent years, climate change activists have pointed out that the proven reserves of fossil fuels on the books of energy companies far exceed the amounts that could be safely burned without causing dangerous levels of climate change. The IEA's *World Energy Outlook* 2012 estimated that we would need to limit CO_2 emissions from energy to 884 billion tonnes in order to have a 50% chance of limiting global warming to 2°C. However, the total of potential carbon emissions from consuming world fossil fuel reserves is more than three times this. In other words, in a world where we meet the agreed climate target, more than two thirds of fossil fuel reserves will need to stay in the ground, at least to 2050. More than two thirds of these reserves, measured by carbon content, are coal. Of course, it is the highest cost and least profitable production that is most at risk.

The Carbon Tracker Initiative (CTI), a non-profit climate organisation, has made a particular study of this topic.[xli] So far they have concentrated on the oil industry. They estimate that if the share of oil in global energy emissions is fixed at the current 40% then the "budget" for oil related emissions to 2050, in a 2°C world, is 360 Gt CO_2. This compares with 635 Gt emissions that would arise from potential production over that period. In other words, many of the investments that energy companies are planning will become "stranded assets" in a 2°C world. Assuming that it is the least cost production that will survive in a low-carbon world, only production that is profitable at a price of $75 per barrel will be viable. They estimate that some $20 trillion of capital expenditure has been earmarked by the oil industry for projects that will require an oil price of $95 per barrel to earn an acceptable return. This includes deep water and Arctic production, and other costly conventional production, as well as oil sands, and heavy oil. 71% of this production would be by wholly private sector companies, which is why CTI are anxious to alert investors to the risks.

These are legitimate questions to raise, which is why the Governor of the Bank of England, Mark Carney, has issued a warning that fossil fuel companies cannot burn all of their reserves if the world is to avoid catastrophic climate change and called for investors to consider the long-term impacts of their decisions.[xlii] Lord Browne, the former CEO of BP, and others, have pointed out that the oil industry faces an "existential threat" from climate change if it does not alter its policies. Most recently, the

---

[xli] Carbon Tracker Initiative, *Carbon Supply Cost Curves: Evaluating Financial Risk to Oil Capital Expenditure*, May 2014.
[xlii] The Guardian, 13 October 2014.

chief economist of BP has argued that the world is facing a long-term oil glut as producers scramble to exploit reserves before fossil fuel demand goes into a decline.[xliii] According to BP, although oil demand is not expected to peak until the mid-2040s, there is twice as much technically recoverable oil available as the world is expected to need by 2050.

Others seek to put pressure on fossil energy companies by pressing investment companies to "divest" their shares. Most famously, the Rockefeller Brothers Fund, built with the profits of their great grandfather's Standard Oil Co, recently announced that it is selling investments in the fossil fuel industry to put pressure on companies that are adding to climate change. A recent study by the Stranded Assets Programme[xliv] has analysed the effectiveness of such divestments.[xlv] They conclude that these campaigns are unlikely to have much direct impact on the value of fossil fuel companies or their ability to raise finance. Coal companies are considered slightly more vulnerable than oil and gas companies. They consider that these campaigns will be at their most effective in "triggering, a process of stigmatisation" of fossil fuel companies. Campaigns may affect the reputation of a business, for its ability to recruit and motivate high quality staff, and how investors see its long-term future. They judge that stigmatisation could cost fossil fuel companies billions, but that it is unlikely to threaten their survival.

The energy industries take a relatively robust line on these threats. The most outspoken, EXXON, describe the divestment movement as, "out of step with reality" and at odds with the need for poor nations to gain better access to energy, as well as the need for fossil fuels to meet global energy demand for decades to come.[xlvi] Shell's line is "we need to temper our expectations of a zero-carbon future because demand for energy is so strong and renewable energy sources are unlikely to be a realistic alternative to fossil

---

[xliii] Financial Times, *BP Warns of Price Pressures from Long-term Oil Glut*, 26 January 2017.

[xliv] The Stranded Assets Programme is a project of the Smith School of Enterprise and the Environment, Oxford University.

[xlv] Ansar, A. Caldecott, B. and Tilbury, J. 2014. *Stranded Assets and the Fossil Fuel Investment Campaign: What Does Divestment Mean for the Valuation of Fossil Fuel Assets*, Stranded Assets Programme, Smith School of Enterprise and the Environment, Oxford University.

[xlvi] National Journal Daily, 14 October 2014.

fuels for many decades.[xlvii] Rio Tinto's climate change position statement of August 2012 recognises that the company, which is a major producer of coal, iron ore, and uranium, has a role to play in addressing climate change, consistent with creating long-term shareholder value. This includes research and development, lower carbon operations, and including carbon prices in investment decision making.[xlviii]

This debate demonstrates the size of the gap between our aspirations for carbon reduction and today's realities. Oil is the fuel of transport throughout the world. We all depend on it. Electric cars are an exciting prospect. Chapter 6 argues that governments should be promoting them strongly. But today they have a very small market share, even in developed nations. In the developed world oil demand is in gradual decline, as the market for cars becomes saturated and vehicle efficiency improves. Hopefully, electric or possibly hydrogen vehicles will begin to contribute significantly. But in the developing world demand for oil is growing rapidly as the new middle classes claim their right to personal mobility. The oil industry is investing to meet this demand while replacing rapidly declining production from existing fields.

The oil companies are an easy target. But demonising them is not going to help. It's as though we pulled up at the petrol station and as the attendant filled up our tank we wagged our finger at him said, "you know you really shouldn't be producing all this oil!". The measures that will curb oil emissions include much tougher regulation of vehicle emissions, investment in public transport, and eventually the switch to electric power. Biofuels can also play a part. We need to spend more on research that can improve the performance of electric vehicles and on advanced technologies for producing oil from plant waste. The oil companies themselves advocate carbon taxes to curb demand and provide a predictable economic basis for the transition to non-oil alternatives. Once these measures are in place, and reflected in the trends for personal and business transport, the oil industry will revise its production plans accordingly.

It is reasonable enough to alert investors in fossil fuel industries to the risks of investing in fossil fuel industries. As mentioned above, BP is

---

[xlvii] Financial Times, 7 December 2014.

[xlviii] Rio Tinto, *Climate Change Position Statement*, August 2012.

already signalling its concerns about oil prices. The coal industry in the US has been placed in serious financial difficulty, mainly because of the shale gas revolution, but also partly because of developing environmental regulation. The coal industry in China also faces mass redundancies because of the economic slowdown and China's efforts to rebalance its economy, partly for environmental reasons. Parts of the oil industry have suffered serious financial difficulty because of the rapid fall in oil prices since 2014, mainly the result of the rapid growth of shale oil production in the US and, again, the slowdown in China. As of June 2017, they are enjoying a partial recovery. Many of the investment projects for relatively high cost oil identified by the CTI have been cancelled already. So it is definitely the case that investments in oil and coal are risky and the possibility of stronger environmental policies which would further reduce demand is now an important additional threat. But boycotting fossil energy investment is not going to contribute significantly to reducing carbon emissions. The oil industry will go into decline when we adopt alternatives, and not before.

15. Conclusion

Climate change is a grave concern which poses a special challenge to mankind's ability to cooperate in the common interest. We have a moral obligation to respond. Efforts at collective action, mainly through the Unite Nations climate treaty, have been impressive but so far they fall far short of what is required. US withdrawal is a disappointment, but the Paris agreement will continue. Today's trends in greenhouse gas emissions are truly alarming. The developed nations have to give a lead. But the core challenge is to find ways in which developing nations can manage their climate emissions while delivering on their top priorities for economic growth and poverty alleviation. It is to be hoped that many nations will, indeed, "ratchet up" the ambitions of their climate policies. But it is now hard to believe that we will succeed in limiting climate change to 2°C unless technological progress opens up new avenues for cost effective emissions reduction. Energy technology, and the possibilities for achieving this, are the subjects of the next chapter.

Chapter 6

Energy Technology

1. Introduction

Carbon emissions from energy threaten the climate and we are far from being on course to meet mitigation targets. In India, Africa, and other developing countries energy shortages are holding back economic and social progress, and even in wealthier countries energy costs are a burden for poorer people. The volatility of fossil fuel prices disrupts the economies of energy producing and consuming countries alike, and especially energy investment. And tensions over access to oil and gas reserves simmer just below the surface of international geopolitics.

Hopefully, in the lifetimes of our grandchildren or great-grandchildren modern energy will be universally affordable from renewable or other clean sources and our present reliance on fossil fuels will appear, with hindsight, costly, dirty, unhealthy, exploitative, and irresponsible in relation to the environment. Perhaps also, the world will be less dependent on international trade in oil, gas, and coal.

There are encouraging signs that we are moving in that direction but we have a long way to go. Today world energy is still dependent on coal, oil, and gas. Change is definitely coming. But the alternatives are, in some cases, still relatively expensive, or are not reliable, and the pace of change is not nearly as rapid as required. As the pace of change accelerates we will increasingly have to find ways to manage the social and geopolitical consequences of the disruption that is caused. But that is a problem that is only just beginning to emerge.

You can be an optimist or a pessimist about the progress of climate mitigation, depending on which end of the telescope you prefer to view. Arguably, shockingly little has changed since the adoption of the climate treaty (UNFCCC) in 1992. The share of fossil fuels in world energy, at over 80%, is much the same as it was. Some of the world's largest and fastest growing economies, including China, India, and Indonesia continue to invest in coal power on a large scale. Global energy related carbon dioxide emissions stalled in 2015 and 2016, but may still be on a gradually rising trend. The new renewables, solar and wind, only accounted for about 2% of world energy in 2016. In the International Energy Agency's (IEAs) estimate, based on the government policies declared for the 2015 Paris climate summit, we are heading for carbon emissions in 2040 of about double the level that would be needed to restrain global warming to 2°C.

On the other hand, some very exciting changes in technology are taking place. The average cost of large scale solar power fell by 65% between 2010 and 2015, and the cost of onshore wind by 30%. Renewables accounted for 70% of total investment in new electricity investment in 2015 and for the first time renewables capacity additions were greater than those of fossil fuels and nuclear combined. This investment was well spread around the world in China, Europe, the US, India, and Brazil. The IEA have commented that, "the trend of rapidly growing deployment of wind, solar PV and (to a lesser extent) hydro over the last decade is unmistakable and perhaps one of the clearest signs of an energy transition taking place".

Further research and deployment at scale may continue to push down the costs of renewables until they become fully competitive with conventional energy. We may also succeed in developing storage and "smart grid" technologies that can deal with the problem of intermittency. So if you are an optimist you may feel that some sort of tipping point is taking place.

Technologies reviewed in this chapter, renewables, nuclear power, and fossil energy with carbon storage, have the potential to produce the energy that we need with very low carbon emissions, especially if we drastically improve the efficiency with which energy is used. But the outcome of the Paris climate summate, where national commitments fell well short of what is needed, strongly suggests that technological advances are necessary, as well as strong political will.

This chapter offers a realistic view of where we stand in the effort to meet global energy needs in an environmentally acceptable way, the contribution that energy technologies can make, and what we need to do to have the best chance of success.

1.1. *Past Experience*

Energy technologies have been extraordinarily long lived. The technologies that we rely on most today, coal powered steam, and the internal combustion engine, have been around for more than a 100 years. The other main sources of energy today — the pressurised water reactor (PWR), the gas turbine, and gravity driven hydro turbines — have all been around for at least 50 years. They have all been greatly refined and there have been big improvements in efficiency. But the basic technologies have proved remarkably robust.

Many energy technologies "of the future" have come and gone. The full implications of deploying new technologies on the vast scale required to make a difference to world energy can be difficult to predict. In the 1970s, nuclear fission was widely expected to become the main source of electricity. But the advent of cheap gas in the US and public concerns following the Chernobyl nuclear disaster have slowed its progress, at least in the West. Nuclear fusion, hydrogen, fuel cells, advanced bioenergy, and fossil fuel with carbon dioxide storage, have all staked their claims to leading roles but have struggled to overcome the technical, social, or economic hurdles to mass adoption. So far, they have not fulfilled expectations. To varying degrees these technologies have found their niches and have become significant contributors. But they are no longer seen as dominant solutions. In considering the outlook for energy technology we must bear in mind that the future is not always what we expect. "The best laid plans of mice and men gang aft aglay".[i]

---

[i]Robert Burns, *To a Mouse*, 1785.

1.2. *Technological Progress*

Most of the progress in energy technology in recent decades has come not from fundamental scientific breakthroughs but in the form of incremental refinements of existing technology and cost reductions associated with mass production and mass deployment. This is true of photovoltaic (PV) solar electricity and it begins to look as though it may be true of batteries and energy storage. Cost reductions due to deployment in one country become available all over the world. Improvements in wind energy have come from increases in size and height, and improvements in the efficiency of fossil plant come from increased temperatures. These gains are linked to steady improvements in materials technology.

The lack of fundamental breakthrough is, perhaps, a disappointment. The improvements that have been derived from refinement and deployment demonstrate that, to a large degree, the future is in our own hands. By investing in advanced technologies we make them more competitive. It is definitely possible to meet the 2°C target through the continuing deployment and incremental development of existing technologies, even if there are no real breakthroughs. But it will require a big increase in political will and financial support. Is this political will available? The headline conclusions of the Paris summit proclaim that it is. The specific national "contributions" that have been offered, reviewed in Chapter 5, suggest probably not.

1.3. *Local Conditions Matter*

One might imagine that the best energy supply options would be the same across the globe. For transport this is largely the case. Oil is highly condensed and easily transportable. The motor industry is global and today the petrol driven internal combustion engine is more or less universal. Refinements that are brought in to meet efficiency regulations in one part of the world spread quickly across the world.

The position is quite different for electric power. Here local conditions make a huge difference. The geography of countries such as Norway and Brazil has made hydropower their main source of electricity. Coal, and to a lesser extent gas, are costly to transport and costs of production vary. In Wyoming and in parts of Canada and Australia, where large out-crops are near the surface, coal is cheap. In these locations, once the overburden has been removed, the coal can be scooped up by huge exca-vators. In other places, for instance most of China, more costly deep underground mining is needed. Coal is relatively plentiful in China, India, and the US but in Europe, except for Poland, most of the accessible, low-cost, reserves have been worked out. Nuclear power is in principle avail-able anywhere in the world provided that the necessary expertise in engineering and management can be assembled. In practice, its availabil-ity is limited by public opinion.

Wind power depends, obviously, on local weather, as does solar power. In southern California, where air conditioning is general, peak demand for electricity broadly coincides with peak sunshine, whereas in Northern Europe peak demand is in the hours of darkness. So Northern Europe will need a lot of storage capacity, or other sources of flexibility, if it is to be largely reliant on solar energy.

In many hot regions, including parts of China, Greece, and South Africa, solar "thermal" collectors are widely used as a source of hot water. This is a highly attractive, low tech, option that can make an increasing contribution to carbon reduction. But it does not generate electricity.

It is reasonable to expect that the energy landscape of the future will be a mosaic, just as it is today. For instance Norway, with its large hydro-power capacity, will find it much easier to switch to renewables than Japan with its large population and relatively limited land area. It will be easier to give up coal in Western Europe, where the most economic reserves have largely been worked out, than in Australia which has large low-cost reserves that are still untouched. Energy policy solutions have to take account of local circumstances.

1.4. *There are Many Options*

The technologies that may contribute in the future are countless. We may free huge amounts of natural gas trapped as "clathrate" in the deep ocean. We may have orbiting collectors of solar power. Nuclear fusion, which is the subject of a major international research effort, may start to deliver affordable electricity. Ocean waves, tides, and currents also have potential. None of these possibilities, or many other areas of research, is to be dismissed, although experience has demonstrated the high hurdle of getting from the workbench to mass deployment. This chapter will concentrate on the technologies that are either in use on a large scale today or seem likely to have significant impact in the next few decades.

1.5. *Hydrogen*

In most of what follows, electricity is assumed to be the main energy vector of the future. That is the general view, which also reflects current trends. However, in many uses, such as powering cars or heating homes, hydrogen offers a possible alternative. Hydrogen can be extracted from water or from the air using almost any form of energy and then transmitted through pipes or in compressed form to deliver energy where it is needed through internal combustion engines, or turbines. Hydrogen can also provide the feedstock for "fuel cells" which produce electricity by chemical means. Stored hydrogen could provide an alternative to electric batteries in vehicles and elsewhere. The product of hydrogen combustion, or use in a fuel cell, is benign water vapour.

The main challenges are cost, as ever, and the need to demonstrate safe transmission and storage. These are not insurmountable, and hydrogen may have its role. The main reason why it seems likely to play second fiddle to electricity is that electric networks already exist, and can be developed incrementally, whereas someone will have to place a big bet on hydrogen and provide a suitable distribution network before hydrogen could really take off.

However, it is too soon to dismiss hydrogen altogether as a main vector for the future. Hydrogen might offer solutions in some areas where electricity has limitations. For instance hydrogen can be stored, and could help to make full use of variable energy sources. It is widely assumed today that, in the future, home heating will be supplied through electric power and heat pumps. But this is an expensive and inconvenient option. Converting domestic gas networks, where they exist, to hydrogen might be a feasible alternative, for instance in the UK, which has an extensive gas network. Hydrogen causes the "embrittlement" of iron pipes, but this may not be such a big problem because many are now being replaced with plastic.

Similarly, hydrogen stored at high pressure offers an alternative to electric batteries in vehicles. If the problem of range anxiety proves difficult to solve, we may need to turn to hydrogen as the low carbon fuel for vehicles. Indeed, the Japanese are aiming to have 40,000 hydrogen fuel cell vehicles on the road, and 160 fuelling stations, by 2020.[ii]

2. Electric Power

One bet that seems fairly safe is that electricity's role will grow. In 2014, electricity contributed about 18% of total energy consumption, and just over 40% of carbon emissions.[iii] It is instantly available, flexible, and clean at the point of use. It has many rapidly growing applications, including in information technology and communications. It's the modern energy source. As discussed below, electricity also seems to provide the most realistic low-carbon alternative to the fossil fuels that are used today for transport and for the heating of buildings.

If electricity is to be the main energy vector of the future, what are the main options for generating electric power?

Figures 6.1 and 6.2 show how our electricity was produced in 2016, and how the IEA project that it may need to be produced in 2040 if we are to meet the environmental target of restricting global warming to 2°C.

We now consider each of these sources in more detail.

2.1. *Coal*

Coal provided 37% of world electric power in 2016, making it much the largest source. It also contributes 46% of global carbon emissions. This is not going to change suddenly. Major emerging economies such as China, India, Indonesia, are largely reliant on coal. That is why demand has been growing rapidly in recent decades, increasing by more than 75% between

---

[ii] Scientific American, *Japan Bets on a Hydrogen-Fueled Future*, 3 May 2016.
[iii] International Energy Agency, *World Energy Outlook*, 2015.

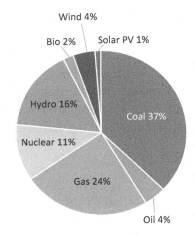

Figure 6.1.   World Electricity Generation in 2016.

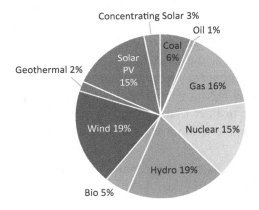

Figure 6.2.   World Electricity Generation in 2040, on 2°C Course.

Source: IEA, World Energy Outlook (2017).

1990 and 2014. This growth has slowed, especially in China, but is continuing. The expected increase, at 0.3% p.a. between 2016 and 2040, is slower than for gas and nuclear power, and much slower than for bioenergy, wind, and solar, so coal's share is gradually declining. Coal is the most polluting of the fossil fuels, and the most productive greenhouse gas.

So to meet the 2°C climate target we will have to find ways to drastically reduce our use of coal, or else tame its emissions.

Coal power technology has been making significant gains. Advances in heat resistant materials, especially nickel alloys, are making it possible to progressively increase steam temperatures and pressures in coal power stations. The latest technology, known as "ultra-supercritical", can achieve efficiencies of 46%, a big improvement on the average efficiency of existing power stations, which is 33%. Greenhouse gases are reduced in proportion. Modernisation of the world's coal power stations would lead to huge emissions reductions, potentially more than 5 billion tonnes of CO_2 p.a. But this poses a dilemma. Even if it were possible to conduct this wholesale upgrading, the new plant would have an operating life of 40–50 years, locking in substantial, albeit somewhat reduced, carbon emissions for the future.

This dilemma is common to many of the questions about how we can best mitigate climate change. If we could be confident that the alternative would be clean energy in one form or another then it would make no sense to promote even the most efficient forms of coal power. But if low-carbon alternatives are not readily available the alternative, in practice, is likely to be less efficient coal or restricted economic and social development due to power restrictions. In that case the right course is not so clear. We must also have due regard to the priorities of the governments of the countries concerned. This is a general topic that will be revisited later.

2.2. Gas

Gas is the next most important source of electric power after coal, and its use continues to grow as additional sources are found around the world. The huge impact of shale gas technology in the US and the transformation of international gas trade through the market in Liquefied Natural Gas (LNG) have been described in Chapter 2. In the IEA's central case, although wind and PV experience the greatest percentage growth rates in the coming decades, in absolute terms it is gas that delivers the largest increase in energy supply. In 2014, the IEA envisaged a "golden age of gas". It has somewhat retreated from that but gas remains a big contributor with a promising future.

As far as gas combustion is concerned, the most significant recent development has been the mass deployment of the combined cycle gas turbine (CCGT) from the late 1980s on. This incorporates many technical improvements, including increased gas temperatures, but the key feature is that two sequential power-generating turbines are used. The first operates from the hot combustion gasses, as in a single cycle turbine, and the second operates from steam generated from the heat of the combustion gases after they have passed through the first turbine. As a result, the efficiency of state of the art gas turbines increased from around 35% at the end of the 1980s to nearly 60% in the late 1990s.

This was a massive shift in the energy technology landscape which, besides increasing the efficiency and reducing the carbon intensity of gas generation, also made gas power much more competitive with coal. The share of gas in power generation in OECD countries more than doubled from 10% to 24% between 1990 and 2014. Emissions are more than a billion tonnes less than if this additional power had come from coal. We would be in a much worse place today on global carbon emissions without CCGT.

Should we regard the development of the CCGT as incremental? It represents the coming together of a series of technological achievements in advanced materials and turbine design, plus the innovation of adding a second turbine — all developments of existing technologies. We should not underestimate what can be achieved through innovations of this kind.

Power from gas has about half the greenhouse gas intensity of coal, provided that its production and transmission is well regulated and managed to avoid leakage to the atmosphere. The extent of such leakage in the US has been a major concern and under the Obama administration, US authorities were in the process of tightening regulation. However, at the time of writing, in June 2017, Trump administration seems set to reverse this progress. Gas is commonly the fuel for Combined Heat and Power (CHP) plant, where the heat from electricity generation is used to provide heat for buildings or industrial processes. This provides a further big improvement in overall efficiency and a reduction in carbon dioxide emissions per unit of useful energy.

In many places, diversification from coal to gas is one of the lowest cost options for reducing carbon emissions. But of course, just like the construction of efficient new coal power stations, the building of new gas stations and infrastructure risks locking in a certain level of carbon emissions for the future. Hence, gas is sometimes referred to as a "transition" technology from the perspective of carbon reduction. As with coal, carbon capture and storage (CCS) is a possible option for reducing the emission from gas power stations in the future.

2.3. *Carbon Capture and Storage (CCS)*

In the longer term, by about 2050, if we are to come anywhere near the 2°C global warming target, we will not be able to rely on conventional coal or gas, as currently used, to any large degree. But there is a technology on the horizon, CCS, that could make it possible to use fossil fuels in a much cleaner way. This involves extracting the carbon dioxide from the exhaust of coal or gas power stations and pumping it into porous underground formations, such as sandstones. There is no doubt as to its technical feasibility. All the stages of this have been demonstrated. Indeed in parts of the US the pumping of carbon dioxide into oil reservoirs to enhance production is a common commercial practice.

There are several technologies available. The carbon dioxide can be extracted from the fuel before combustion, or the fuel can be burned in an oxygen environment to produce an almost pure carbon dioxide exhaust stream, or the carbon can be extracted after combustion. They all require a large increase in the original capital cost of the power plant, as well as higher operating costs. The energy used to extract and then compress the carbon dioxide for transport causes a significant reduction in the overall efficiency of the power station. So CCS, in contrast to other low-carbon technologies, leads to a *reduction* in useful energy supplied. There are some circumstances where the carbon dioxide produced has a positive value, for instance in enhanced oil production, as I have mentioned. At present, the potential for that is limited. Researchers are looking for other uses of carbon dioxide. It has many uses in industry and, indeed in food. But unfortunately none of these is on anything like the right scale. Without a commercial use that gives positive value to carbon dioxide, CCS can never be competitive unless it is directly subsidised or unless there is a charge, or other restriction, on carbon dioxide emissions.

CCS also faces difficulties of public acceptability. The carbon dioxide has to be pumped at high pressure to the reservoir, sometimes a long distance, and then injected. How safe is all this? How can we be certain that the carbon dioxide will remain in place in the reservoir and will never leak out?

Regulatory regimes for managing these risks are in preparation in the UK and elsewhere. Experience so far shows that, in the right geological conditions, carbon dioxide can be stored successfully underground and is stable over very long periods. Because of the complexity of underground formations it is difficult to prove that there could never be any leakage. But the possibility that some small fraction of the CO_2 might leak out over thousands of years is hardly a big problem when one considers that the alternative may be that all the CO_2 will be vented straight into the atmosphere! Nevertheless, convincing the public that the risks are acceptably low is another matter. In Germany, for instance, demonstration of large-scale onshore storage has ground to a halt because of public concerns.

The drawbacks are formidable, but many modelling studies indicate that the successful deployment of CCS is essential if we are to meet the 2°C target. This is partly because coal and gas are so dominant in power generation around the world. In some circumstances, even with the additional cost and loss of efficiency of CCS, they may still provide the lowest cost and most accessible form of reliable low carbon power. And the political difficulties of closing down the coal-mining industries are formidable. Also, many industrial processes, such as iron, steel and cement manufacture, require very high temperatures or chemical input that are difficult to achieve without fossil fuels. They are amongst the largest sources of carbon emissions.

As already mentioned, there is no doubt about the technical feasibility of CCS. However, we won't really know the true cost or whether it can win public acceptance until we press ahead with a succession of full-scale demonstration plants. Unfortunately governments round the world have dragged their feet on this. In 2005, ministers of Western governments agreed at the IEA that 10 full-scale projects would be under way by 2010. Nothing of the sort has happened. It is fair to say that progress has been sclerotic. After more than 10 years of promising to support several full-scale CCs projects, and two rounds of proposals by industry, the UK government finally pulled the plug on full-scale implementation of CCS in 2015.

There are a number of plants around the world in which carbon dioxide is extracted from the exhaust of various kinds of industrial plant and used to enhance production from local oil fields. But these are not major power stations. There are more projects at various stages of development, mainly in the US and China.

Two projects in an advanced stage in the US give an idea of the state of progress. The Kemper County Energy Facility in Mississippi was to be a full scale coal fired power station that used "pre-combustion" technology for capturing the carbon. The coal, or rather lignite in this case, was first to be gasified into a "syngas" that consisted mainly of CO_2 and hydrogen. The CO_2 and other impurities were then to be extracted from the gas, after which the hydrogen was to be burned in combined cycle turbines to produce electricity. This is an advanced, potentially highly efficient, system which may represent the future of CCS. The plant suffered major delays and big cost over-runs that tripled the original cost estimate of $2 billion. Finally, the promoters ran out of patience, abandoned the CCS technology, and decided to run the plant simply as a gas power station. A sad end to the story. This does not rule out the possibility that this advanced and potentially highly efficient CCS technology may be important for the future.

In contrast, a scheme for retrofitting more conventional CCS technology to the Petra Nova generating station in Texas, has just come into operation on schedule and within budget. This is a "post combustion" project in which the CO_2 is extracted from the power station exhaust. The project is costing a relatively modest $1 billion. This does not advance the technology as much as Kemper County would have done, but it demonstrates the possibility of containing the costs of well managed CCS projects. Both projects have had some support from the US government, but they are mainly financed by big utilities. In the absence of carbon taxes, CCS projects will have to earn their keep through the use of CO_2 to enhance recovery from oil fields in the region.

There is another reason to take CCS seriously. The 2°C target may not be low enough. It will not prevent significant climate hazards or the threat of inundation to small island states. That is why, at the Paris climate summit, world leaders agreed to "pursue efforts" to limit global warming to 1.5°C. It is too late to achieve this by reducing emissions. The only credible options will require pumping carbon dioxide from the atmosphere back underground. We could take this gas directly from the air, but that is expensive because the concentrations are very low. A much more promising option is to capture and store the carbon dioxide from a power station or industrial plant running on biological material — which achieves "negative emissions" by sequestrating the gas that plants have captured from the atmosphere.

Environmentalists have mixed views about CCS. For instance, the prominent US environmental lobby group, the Sierra Club has fought against the Kemper County project. It is viewed as an "end of pipe" technology which fails to address the fundamental problems of the extractive industries and provides less than complete carbon reduction. It is true that current CCS projects only extract a maximum of around 90% of the carbon dioxide and in most cases much less. If CCS is to have a big part in very low-carbon energy future, further technical progress will be needed. But if we don't pursue CCS we are placing a huge bet that it will be practical to virtually eliminate coal and gas from energy supply within three or four decades.

There are serious drawbacks to CCS. But, along with nuclear power, bioenergy, and hydropower, it is one of the few options available for producing reliable, low-carbon, baseline electric power. At present, it appears the only option for decarbonising heavy industry.

2.4. *Nuclear Power*

Nuclear power is a substantial contributor to low-carbon power today, second only to hydro. It supplies 11% of electricity worldwide and this is expected to increase gradually to 2040. In some countries it is dominant. For instance nuclear power is the main source of electricity in France and, as a result, France has about half the carbon emissions per head of the OECD as a whole.

The peak of investment in nuclear power in the West was in the mid-1980s. Around 30 GW p.a. of new nuclear capacity was commissioned in 1984 and 1985, mainly in Europe and the US.[iv] But a decline in oil and gas prices and difficulties containing nuclear costs, together with a serious nuclear scare at Three Mile Island in the US in 1979, and a full-blown

---

[iv] International Energy Agency and Nuclear Energy Agency, *Nuclear Energy Technology Roadmap*, 2015.

disaster at Chernobyl in Ukraine in 1886, have since led to a progressive reduction in new nuclear orders, at least in the West. However in 2015, China had more than 20 nuclear power stations under construction, and there were other major construction programmes in Russia, Korea, and the UAE.[v]

Most of the costs of nuclear power are in the original cost of building the station. Individual stations tend to be very large. For investors to get a reasonable return the station must be allowed to operate at high capacity for 20 or 30 years and the power it generates must be competitive with other sources over that period. Nuclear power is not particularly well suited to "load following" to meet short-term variations in demand but in France, where it plays the largest role, a considerable degree of flexibility has been demonstrated.

Historically, energy companies that built nuclear power stations in the 1970s, when energy prices were high, went bankrupt when cheaper oil and gas came into the picture. They played a major part in helping to break the OPEC cartel but then they were ruined when it collapsed. The "chunky" nature of nuclear investment, in which very large sums have to be committed up front for a single project, is a serious problem. In contrast, it is a big plus for renewables such as onshore wind and PV that they come in modular form and offer more manageable investment opportunities.

The latest European nuclear power design, Areva's European Pressurised Reactor (EPR), is in trouble. Three stations are under construction, in France, Finland, and China. Both the stations in France and Finland have suffered major construction delays and problems with steel quality. As of 2016, the French station is 6 years behind schedule and $7 billion over budget. The Finnish station is 9 years delayed and $5 billion over budget. Only the China project is making better progress.

Why are these EPRs so difficult to build? Part of the problem is that new nuclear construction has been suspended in Europe for the past 15 years so there is a lack of experienced people and component manufacturing capability. But the design itself also contributes. The EPR, like most nuclear stations around the world today, is a variant of the PWR originally

---

[v] International Energy Agency, *Energy Technology Perspectives*, 2016.

created to power US submarines. For this purpose it needed a very compact core. However, there are obvious problems in depending, to cool and control the reactor, on very hot water that is only prevented from boiling by being kept under high pressure in a steel container. If for any reason the pressure vessel is fractured, or the supply of water is cut off, the reactor can run out of control very quickly. This is what happened at Fukushima in 2011, when a tidal wave swamped the pumps delivering water to the reactor, and what happened at Three Mile Island, in the US, in 1979.

I remember, when I was working for the British Government back in the 1980s, attending a meeting with the French designers of the EPR. They were working in partnership with the Germans and they complained that their German colleagues were somewhat over the top on safety criteria. The French recognised that the reactor needed to be robust against an aircraft crash. However, the Germans were insisting that it should be capable of withstanding the impact of two Boeing 747s, the largest aircraft of the time. At the time I remember thinking that this was rather fanciful, but that was before 9/11!

The nuclear components of the EPR are contained in a huge reinforced concrete shell that is lined with steel. There are four separate core cooling systems, and the base of the reactor includes a "catcher" for managing any leakage from the core. It's a very safe system. But it seems to be a devil to construct.[vi]

Other reactor types besides the EPR that are under consideration for construction in the UK are the Chinese Hualong (HPR-1000), Toshiba Westinghouse's AP 1000, and Hitache-GE's Advanced Boiling Water Reactor (ABWR).

In Europe, new nuclear stations are experiencing major cost overruns. In order to secure the UK's first EPR, at Hinkley Point C, the UK government has had to guarantee a subsidised price for the electricity, calculated to yield an adequate return on the $25 billion of investment required. In the US the Energy Information Administration (EIA) estimates the cost of electricity from new nuclear plant, coming into service in 2022, at almost twice the cost of combined cycle gas or onshore wind, which, except for

---

[vi] Financial Times, *EDF's Nuclear Troubles Rooted in Caution*, 20 March 2016.

geothermal, are the lowest cost sources.[vii] In China and India, nuclear power is competitive with other sources of generation and plays an important part in national energy plans.

Nuclear power has been dogged by concerns about safety, waste management, and the proliferation of radioactive materials. Public acceptance is its Achilles heel, at least in the West. For years any attempt at rational public debate on energy options has been dogged by fierce arguments between those who are for and against it. In Austria and Italy, nuclear power is banned. In Germany, it is being phased out. In Japan, following the 2011 Fukushima disaster, all nuclear plants were shut down. In May 2017, the process of restarting some of the reactors has begun. But public concerns may still limit the process of recovery. In the meanwhile, the government has achieved remarkable reductions in electricity demand, discussed in Chapter 8 on energy efficiency, but has had to rely on high cost distress purchases of natural gas. But some of the countries with the largest energy needs, including China, India, Russia, and Korea, are pressing ahead with major programmes of new stations.

Statistically nuclear power is very safe, but when accidents do occur, they rank as major disasters, especially at Chernobyl which has been by far the most serious. The insurance of nuclear power stations depends on government support. The Paris Convention on Third Party Liability in the Field of Nuclear Energy (1960) and Brussels Supplementary Convention (1964) embody complex international agreements on this topic.

Many people have a particular horror of nuclear contamination, even at very low levels. There is particular concern about the management of nuclear waste. The issue of permanent disposal of high-level nuclear waste came to a head in the US, and received widespread publicity, when, in 1987, the US government designated Yukka Mountain in Nevada to contain its waste depository. The project became a focus for environmentalists and those opposed to nuclear power and after continuing outcry it was finally dropped in 2011.

[vii] US Energy Information Administration, Levelized Cost and Levelized Avoided Cost of New generation Resources in the *Annual Energy Outlook 2016*, August 2016.

No country yet has a permanent geological high-level nuclear waste depository in operation. Finland has announced the site for its repository. Sweden and France are working towards their depositories. Successive UK governments have been looking for geological sites to dispose of nuclear waste for many decades, but all efforts have foundered against local resistance. The latest plan, set out in July 2014, is based on a "voluntarist approach", in the belief that a Geological Disposal Facility (GDF) will confer "significant economic benefits" on a locality that accepts it.[viii] However, no volunteers have come forward so far.

In the meanwhile, waste is stored above ground. For instance, the new power station at Hinkley Point C will have an Interim Spent Fuel Store that is expected to retain the fuel for at least 100 years. The UK has a massive legacy of nuclear waste from its early nuclear power and weapons programmes. This presents a much more serious management problem than used fuel from a new nuclear station that will be designed with long-term storage in mind.

Many countries have adopted the principle that the waste, even when consigned to its underground store, should be capable of being retrieved in the future, should this be judged desirable for any reason.

Nuclear waste is an emotive topic and some regard the lack of final solutions to the problem of high-level nuclear waste disposal as a strong reason not to invest in new nuclear power. Certainly this is a problem that has to be managed very carefully. However, the world is full of problems that don't have final solutions. Safe interim management of used nuclear fuel followed by eventual interment in long-term depositories seems like the best available option, and that is being widely adopted. The burden for future generations, if any, should be modest. Certainly much less than the threat of climate change.

Nuclear power is associated with nuclear weapons and another concern is that as nuclear power spreads across the globe, so does the risk that rogue regimes or terrorists will get access to nuclear materials. There is a UN agency, the International Atomic Energy Agency, dedicated to minimising the risk of nuclear proliferation and to the inspection of nuclear facilities. But no arrangements of this kind can be perfect.

[viii] DECC, *Implementing Geological Disposal*, July 2014.

Advanced Nuclear Reactors

Advanced nuclear reactor designs, known as "Generation IV", are under development around the world, especially in the US, China, India, France, and Russia.[ix] The broad aims of Generation IV are:

- Improved economics,
- Enhanced safety,
- Minimal waste,
- Proliferation resistance.

There are nearly 100 different Generation IV concepts. Six of these have been identified by the Gen IV International Forum as being of the greatest interest. They are:

- Gas cooled fast reactor (GFR),
- Lead-cooled fast reactor (LFR),
- Molten salt reactor (MSR),
- Sodium-cooled fast reactor (SCR),
- Supercritical-water-cooled reactor (SCWR),
- Very-high-temperature reactor (VHTR).

Of these, it is the GFR and the VHTR concepts that are the most advanced and are receiving the most attention.

The SFR has two main potential advantages over the PWR or BWR. The first is that the coolant, sodium, has a much higher heat density and boiling point than water, making the reactor both more efficient and safer. The second is that the SFR largely recycles its own fuel, originally seen mainly as a means of extending fuel supply, but now valued more as a means of minimising, or even eliminating, the need for high-level waste disposal.

The main advantage of the VHTR is its exceptional safety. The large graphite core gives the reactor exceptional heat stability and there is a

---

[ix] GEB IV International Forum, *Technology Road Map Update for Generation IV Nuclear Energy Systems*, January 2014.

strong negative temperature coefficient of reactivity; in other words if the reactor heats up this causes the nuclear reaction to slow. The VHTR is also well adapted to provide heat for heavy industrial processes and also to produce hydrogen, in addition to the power generation which is usually the main purpose of civil nuclear reactors.

The timetable for implementing all the Gen IV technologies has slipped in the past decade. The validation of the design and performance of the SFR and VHTR concept are at an advanced stage. The next stage, which could start during the 2020s, will be the licensing, construction, and operation of a full-scale demonstration reactor. This phase is expected to last for at least 10 years and to cost several billion dollars.

The sheer size of full scale nuclear power stations presents a problem, at least in the West where most stations are expected to be privately financed. £15–20 billion is a very large bet to place on a single project, especially where there have been construction problems with the design elsewhere, and financing is inevitably difficult. The scale of onsite construction and assembly is difficult, as the problems have shown. And even if construction, goes reasonably smoothly, interest costs mount up substantially during construction. For these reasons, interest has been growing in the idea of smaller nuclear reactors that would not represent such daunting unit risk, could be factory made and easier to assemble onsite. A number of consortia around the world are working on smaller reactor designs, though none is yet ready for commercial deployment.

In the 1970s, when nuclear power was seen as the future of energy, the possibility that we might run out of uranium was a serious worry. But today, as nuclear power is finding its niche as one of a number of major energy sources, and since we have discovered a lot more uranium in the ground, especially in Canada and Australia, this much is less of a concern. This also means that there is less interest than there was in switching to thorium as a nuclear fuel. Thorium is plentiful, especially in India, and could vastly increase nuclear capacity. It also has some advantages over uranium in terms of waste management and proliferation resistance. But switching to thorium requires a major rethink of reactor design and it is not high on the agenda of the nuclear industry today except perhaps in India.

Public opinion on nuclear power is a complex topic. As a general rule, the public all over the world will express a preference for renewables over nuclear power. They also express varying degrees of concern with regard to the safety, waste management and security of nuclear power. These concerns were heightened in the aftermath of Fukushima. A global poll taken by IPSOS just after Fukushima showed only 38% support for nuclear power, down from 54% earlier.[x] But 40% of those opposed supported modernisation of electricity with existing or new nuclear stations. Much depends on timing and on how the question is put.

In Germany, Austria, Italy, and for the time being, at least, in Japan, public opposition has been politically decisive.[xi] In a 2015 Gallup poll, 51% of Americans said that they somewhat or strongly supported nuclear energy, compared to 41% who opposed it.[xii] A 2014 study of attitudes in the UK found that public opinion in the UK was also evenly balanced between those supporting or opposing nuclear power, and that the level of support had remained surprisingly robust after the Fukushima accident.[xiii] Here again, the responses depend very much on how the question is put. There are clear majorities willing to support nuclear power if it is needed for security of supply or to address climate change. People who have got used to living near nuclear installations are generally more favourable than others.

In the UK the debate around the first new nuclear order for several decades centres on the economics and the energy policy context, rather than safety or environmental concerns. In China there is much higher degree of willingness to trust the government on nuclear power issues than in the West. However, there is a growing level of public concern.[xiv] In 2013, public protests led to the cancellation of a planned uranium processing facility in Guangdong province. Probably, this will lead to improved public information and consultation in the future. Public concern has also been growing in India, and there have been mass public protests to proposals to build new stations in Maharashtra and Tamil Nadu.

[x] Ipsos Global advisory, *Sharp World Wide Drop in Support for Nuclear Energy as 26% of New Opponents say Fukushima Drove their Decision.* Ipsos, Monday 20 June, 2011.

[xi] Guizhen, H. *et al.* Nuclear Power in China after Fukushima: Understanding Public Knowledge Attitudes and Trust. *Journal of Risk Research* 2012, 1–17.

[xii] Riffkin R, *Support for Nuclear Energy at 51%.* Gallup press release, 30 March 2015.

[xiii] UK Energy Research Centre (UKERC), Synthesis Report, *Public Attitudes to Nuclear Power and Climate Change in Britain 2 Years After the Fukushima Accident,* 2014.

[xiv] Guizen, H. *et al. Ibid.*

Recent experience, at least in the West, has seen the cost of nuclear power rising, with significant over-runs of new nuclear projects, whereas the costs of renewables continue to decline. Is this a reason to give up on nuclear power? The story of rising nuclear costs bears closer examination. It is certainly true that we have seen costs increase, especially in the US and to some extent in France, during periods of regulatory review when construction has been halted and regulatory requirements changed. The new EPR is expensive for the reasons that I have explained. But there are also examples, for instance in Korea, which demonstrate that settled and sustained nuclear programmes with a high degree of design stability can benefit from significant cost reductions.[xv,xvi]

So a critical question for nuclear power is this. Is it inherent in the risks associated with nuclear power that governments and regulators will continually demand additional and costly safety features? Alternatively, will it be possible to settle on one very safe design, such as the EPR, for a continuing programme of new stations thus, at last, benefiting from standardisation? Failing that, will the next generation of designs be so inherently safe that the continuing search for additional safety feature become outdated?

Undoubtedly, legitimate public concern is a challenge to the future of the industry all around the world. However, experience in the US and the UK suggests that the public may accept nuclear power provided that it is well regulated and managed, that there is good information and consultation, and, most important of all, that there is a convincing case that it is needed.

If we had alternative sources of constant low-carbon energy available everywhere, and if we were comfortably on course to deal with global warming, then we would have the luxury of doing without nuclear power. But of course, we haven't and we aren't. Does new nuclear power represent such a threat that we should be prepared to give up one of our main low-carbon options? I don't believe so. Public opinion may come to the

---

[xv] Lovering, J.R. *et al.* Historical Construction Costs of Global Nuclear Power Reactors, *Energy Policy*, **91** (2016) 371–382.

[xvi] Berthelemy, M. *et al.* Nuclear Reactor's Construction Costs: The Role of Lead-Time, Standardisation, and Technical Progress. *Energy Policy*, **82** (2015) 118–130.

same conclusion, at least in parts of the world where nuclear power is most needed. As discussed below, we may find cost effective ways to make our electric power systems so flexible that we can rely almost entirely on variable renewables. That would call in question the need for nuclear power. But we are a long way from having demonstrated that today.

Nuclear Fusion

As I have explained, nuclear *fission*, the splitting of large atoms, plays a substantial role in energy supply today. The story of nuclear *fusion*, the fusing of small atoms, is very different. In principle the great international nuclear fusion project, the International Thermonuclear Experimental Reactor (ITER), is admirable. Scientists and technologists from all over the world have come together to capture the power of the sun for mankind. Nuclear fusion could eventually provide a virtually unlimited source of power that is carbon free and which does not pose the same safety and security issues as nuclear fission.

But delivering these benefits in practice is another story. Fusion, in the form that it is currently being pursued, requires the management of a "plasma" that is so hot that it can only be contained by magnetic forces. Fusion research dates back to the 1950s and large scale test machines have been operating since the 1970s in Russia, the US, and the UK. But we are still nowhere near having a useful power plant. Today ITER is constructing a huge test machine at Cadarache in France. China, the EU, Russia, Japan, and South Korea are all participants. After substantial over-runs it is now expected to cost between €18 and €22 billion. It is not now expected to produce full power before 2035.[xvii] But this is only a test reactor. Building the first prototype power station would be the next stage. It is impossible to tell whether such a power station would be competitive with other sources of low-carbon energy at the time. Perhaps fusion will be the ultimate solution to our energy problems. Or perhaps the governments who are backing it will run out of patience or money. However, for the next few decades we are going to have to find other solutions to our energy challenges.

[xvii] Reuters Commodities, *Nuclear Fusion Reactor ITER's Construction Accelerates as Cost Estimates Swell*, October, 2016.

2.5. *Hydropower*

Hydropower was by far the largest source of low-carbon renewable electric power in 2016, accounting for 16% of world electricity. Obviously it depends on suitable geography. In Norway and Brazil, it is the main source of electricity. In both China and India, hydro is the next largest source of electricity after coal. But the potential for growth is limited by the sites available. Over the decade leading to 2013, hydropower provided more new capacity than any other renewable energy technology, 60% of it in China.[xviii]

Unlike some other renewables, hydropower is dependable over extended periods, although it ultimately requires adequacy of rainfall. In 2015, India suffered a shortage of hydro power as a result of two poor monsoons. Brazil also had diminished hydropower as a result of poor rainfall. Where large reservoirs have been created they provide energy storage and this makes hydropower the perfect complement to variable renewable energy. Norwegian hydropower is an excellent complement for Danish wind power. In the future climate change could affect rainfall patterns, and therefore the availability of hydropower in ways that are hard to predict.

Large areas have to be flooded to create hydro reservoirs and this is usually controversial, especially if people have to be moved. More than a million people had to be relocated in the 1990s for China's huge Three Gorges hydropower station in Hubei province. This is the largest power station in the world, with a generating capacity of 22.5 billion watts (GW), equivalent to more than one third of the UK's total power capacity. Because of these environmental and social drawbacks some more recent hydro schemes, known as "run of river", have been built, for instance in Brazil, without substantial upstream reservoirs and these do not have the same storage potential.

Dam building can also be a source of international tensions. For instance the 2,700 mile Mekong river, which runs through China, Laos, Myanmar, Thailand, Cambodia, and Vietnam, is the largest inland fishery in the world. It is a vital source of water and fertilising silt. The Mekong Agreement

---

[xviii] International Energy Agency, *World Energy Outlook*, 2015.

between Cambodia, Laos, Thailand, and Vietnam, is supposed to promote consultation. But in practice disputes are rife, especially over dams that China and Laos are building in the upper reaches.

2.6. *Geothermal Power*

Geothermal power, extracting energy from naturally hot strata below the earth's surface, is another source that is linked to geography. It has the greatest promise in areas where the earth's continental plates meet, such as around the Pacific, in the Great Rift Valley in Africa, in Iceland and Turkey.[xix] In the right conditions, it is already a competitive source of power. In Iceland, famous for its hot water geysers, about 25% of electricity was from geothermal energy in 2015. However, geothermal accounted for less than 1% of world electricity in 2016. This could increase significantly but, at the moment, it looks as though geothermal, while important in a few countries, will remain a bit player in world energy.

2.7. *BioEnergy*

Bioenergy, the burning of biological materials such as wood pellets and agricultural or municipal waste in power stations, accounted for about 2% of world power generation in 2016 and, taking the IEA's central projection, this could increase to around 3% in 2040. In the European Union, it accounted for 6% of generation in 2016. However, bioenergy is not just, or even mainly, used for power generation. It is worth digressing for a moment to consider its overall impact.

Bioenergy is by far the world's largest source of renewable energy, accounting for 10% of world energy of all kinds in 2016, whereas hydro accounted for 3% and all other renewables, including wind and solar, for just 2%.

However, more than half of bioenergy comes from wood, charcoal, or animal dung used for heating or cooking in the course of traditional lifestyles in developing countries. 50% of energy in Africa is from renewables, mostly of this kind. Often, especially with growing populations, this

---

[xix] International Energy Agency, *Energy Technology Perspectives*, 2008.

exceeds the sustainable capacity of local woodlands. It tends to mean a hard life for women. Indoor air pollution from burning these fuels has created one of the most severe global health crises, especially affecting women and children. There are many worthwhile initiatives to bring modern energy supply, and especially healthier cooking facilities to these people. Hopefully, this energy sector is in decline.

There are many sources of modern bioenergy. They include oil crops such as oil palms, sunflowers, and rape, as well as sugar and starch crops, such as sugar cane and beet. They also include wood and parts of plants that are not suitable for food, including sugar cane residue (known as bagasse) and straw. And they can be derived from municipal or other wastes.

The main uses of modern biofuels are for transport, or power generation.

Blending biofuels, in the form of ethanol or biodiesel, with conventional vehicle fuels is one of the few currently available options, along with efficiency improvements and changes in travel habits, for reducing the carbon footprint of travel, at least until electric or hydrogen power comes to occupy a significant part of the market.

Most developed countries have obligations that effectively require this blending. The EU has a requirement for 10% renewables and the US has a requirement for 9%, specifically of biofuels. Ramblers in the UK will have noticed the rather startling appearance of bright yellow fields of rape which are the consequence of the national 4.75% biofuels target. Political support for these obligations is cemented, especially in the US, by the fact that they provide support for farming as well as reducing carbon emissions.

Brazil is the golden boy of biofuels for transport, largely based on the conversion of sugar cane waste. Most cars can run happily on fuel containing up to 10% ethanol, but many cars in Brazil are Flexible Fuel Vehicles (FFV) that have been adapted to run on up to 85% ethanol. The national blending requirement, as of 2015, is 27%, the highest in the world, and Brazil is also a substantial exporter of ethanol. Indonesia, which is a major producer of palm oil, is aiming for 30% by 2025.

The IEA see an important role for advanced biofuels in limiting the growth of emissions from transport to 2040. This is partly because the alternative low-carbon options for trucks, aeroplanes, and ships are limited. Even for light vehicles they expect that the cost of batteries will slow the penetration of electric power. In their scenario that meets the 2°C climate target, biofuels supply more than 10% of liquid fuels in 2040.

There is a very wide range of technologies generating heat or electricity from biomaterials. Some are very simple such as wood stoves. Others are highly sophisticated. They include anaerobic digestion of wastes to produce methane, extraction of methane from landfill sites, municipal waste combustion, and gasification technologies. On a larger scale, coal power stations can be adapted to run partly or even wholly on wood pellets, a measure that has been adopted for some power stations in the UK as the government strives to meet its tough renewables target.

Germany's pioneering adoption of wind and solar power is well known. What is not so well known is that bioenergy also plays a big part in Germany's low-carbon transition, currently contributing more electricity than solar and almost as much as wind.

New technology has the potential to widen our bioenergy options. Advanced biorefineries can process the woody, or non-food, part of plants to produce a wide range of fuels and feedstocks. This technology has not taken off as rapidly as was hoped at the turn of the 21st century, but there are now a number of biorefineries operating in the US and, especially if oil prices rise, there is the potential for a growing industry.

In principle, modern bioenergy has many attractive features. In contrast to solar and wind energy, it is a reliable source of power able to respond to demand. It can provide fuel for aircraft and heavy vehicles, for which few other low-carbon options are in sight. Biofuel power stations whose carbon emissions are captured and stored, provide the lowest cost option for extracting CO_2 from the atmosphere, if that should eventually become necessary. Biofuels can boost farming. Biofuel crops tend to grow fastest in tropical regions and that means that they provide potential for economic development, import substitution, and exports in some of the poorest parts of the world. Rich countries that buy their biofuels from the South may be contributing to economic welfare as well as carbon reduction.

But biofuels are also highly controversial.[xx] In principle, the cycle in which plants extract CO_2 from the atmosphere as they grow and this

---

[xx] Brack, D. *Woody Biomass for Power and Heat: Impact on Climate Change*: Chatham House research paper, February 2017. Also, *REA Response to Chatham House report-Woody Biomass for Power and Heat*. Renewable Energy Association February 2017.

is then released back into the atmosphere when they are burnt as fuel should be carbon neutral. But in practice, it isn't quite so simple. This is because land use, including forestry, is the next most important sector for climate change, after energy. The question arises, what would have been the alternative use for land that is devoted to growing biofuels? If rain forests have been cut down to make way for biofuel plantations that is plainly not environmentally benign. Most people accept that the use of offcuts from the timber industry is fine. But what about managed forests in which some or all of the timber is devoted to bioenergy?

It's not only carbon emissions that are of concern, because large biofuel plantations may reduce the availability of land for food crops and draw down local water supplies. Some are concerned that large scale "monoculture" of a single biofuel crop can reduce biodiversity. Some of the processes for converting plant material into ethanol are quite energy intensive and, especially if they are coal powered, may significantly erode the low-carbon credentials of the product.

Plainly it would be great if we could establish sound and universally accepted standards that would enable trade and investment in properly certified sustainable renewables to flourish. A lot of hard work has gone into this. The Global Bioenergy Partnership (GBEP) is the leading international institution. The International Organisation for Standardisation (ISO), as well as US and EU authorities have all developed relevant standards. But their validity is hotly contested. A lot of interests are involved and the issues are complex. Developing countries are not necessarily happy to have their land use policies dictated by the developed countries who are the customers for their biofuel exports.

Biofuels have the potential to make an important contribution to solving our energy problems and, in some countries to economic development. We must keep the door open for that. There are legitimate concerns that not all biofuels are equal and some may do more harm to our environment than good. The topic is highly charged politically and technically. It is going to take a long time to develop internationally agreed standards, if that is ever possible. But we should keep trying. In the meanwhile, governments with biofuels as a significant part of their energy mix have an obligation to take a responsible attitude on how they are sourced,

2.8. *Wind Power*

Wind power has started to make a big impact on world energy over the past 5 years or so. It is a big success story. Capacity has increased about five times over since 2005 and global investment in 2014 was a staggering $270 billion. Wind contributed about 4% of global power generation in 2016 and this is expected by the IEA to increase to 11% by 2040, making wind by far the most important of the new renewables over this period. Wind power is, of course, intermittent. According to the IEA, the availability of onshore wind varies from 20% to 49% and offshore wind from 30% to 48%.[xxi]

The costs of onshore wind power depend on local conditions and transmission costs. Overall, costs have been declining significantly in recent years as turbines have become larger and there have been improvements in availability. The US EIA expects onshore wind energy to be more expensive than conventional gas power in 2019, but cheaper than gas power with CCS. The IEA estimate that by 2035 onshore wind may be cheaper than wholesale electricity prices in Europe and about equal to wholesale prices in China. In the US, where wholesale electricity prices are lower, wind may still be significantly more costly.

Offshore wind costs more to install than onshore wind. These costs are partially offset by better wind conditions and, because offshore wind is still relatively new, there may be more potential for cost reductions in the future. Nevertheless, offshore wind has been regarded as a relatively expensive source of electricity. In the cost projections given below it costs around twice as much as onshore wind and, still, significantly more than conventional power. But this may be changing. The latest European offshore wind auctions, in 2017, have seen some spectacularly low prices that may lead to a reappraisal of the cost of offshore wind.

[xxi] International Energy Agency and Nuclear Energy Agency, *Projected Costs of Generating Electricity*, 2015 edition.

In parts of the UK, especially in areas of dense population where the countryside is protected by strict planning rules, there is growing local and political resistance to new onshore wind developments. As a result, the UK is becoming a world leader in offshore wind. In other parts of the world the availability of onshore sites may be less of a constraint, although local opposition movements have also arisen in some parts of the US. One of the factors that have made wind energy popular in Germany is that local landowners or local communities have shared in the revenues.

2.9. *Solar Energy*

PV, which converts sunlight directly into electric power, is perhaps the most exciting of the new energy technologies. This was an expensive technology, but the costs have been declining spectacularly in recent years and this trend seems to be continuing. In 2016 solar PV capacity increased by 50%, rising faster than that of any other fuel and, for the first time, surpassing the net growth in coal. Most of this increased capacity was in China and the US. It represents a truly remarkable rate of progress. In some auctions prices as low as $30 per MWh are being quoted.

The CEO of Shell has said that he expects PV to become the dominant backbone of the world's energy system in years to come.[xxii] This seems a reasonable bet although, as pointed out earlier in the Chapter, it's a precarious business making long-term predictions about energy technologies. As discussed below, the potential of PV is to some extent linked to developments in energy storage technologies.

There are a number of technical options for PV collectors but today the most commonly used material is silicon. Since silicon is the second most abundant material in the earth's crust, there seems to be no overriding reason why PV could not be deployed on a very large scale in the future. The US EIA estimate that it will cost more than conventional gas with CCS but less than offshore wind in 2019. By 2050, the IEA judges, the costs of PV in suitable locations might be similar to those of conventional plant, such as coal and gas.

[xxii] CleanTechnica, September 2015. Available at http://cleantechnica.com/2015/09/30/shell-ceo-solar-energy-backbone-worlds-energy-system/. Accessed 26 August 2016.

PV is starting from a low base. In 2016, it accounted for only about 1% of world power generation but, according to the IEA's central scenario,[xxiii] it could account for 8% in 2040. In their longer term projections the IEA see PV as playing a major role in a low-carbon future. PV is, obviously, an intermittent power source only available in daylight hours and varying, to some extent, with cloud cover. I will come back to the implications of this variability at greater length later on.

Besides PV, the, most promising option for generating electricity from the sun is Concentrated Solar Power (CSP) in which the sun's rays are concentrated by reflectors onto a boiler which powers steam turbines in much the same way as conventional power stations. There are many varieties of CSP designs. Deployment is at a much earlier stage and eventual costs are more difficult to judge. However, some varieties of CSP have the potentially important advantage of including a reservoir of very hot liquid that provides a form of power storage.

In the IEA's low-carbon scenario, in which the world meets the 2°C target, solar power contributes 15% of electricity supply by 2040. They do not rule out the possibility that solar energy will be the dominant source of electricity by 2050, contributing 27% of global power.

Obviously hot and sunny regions are the most suitable for solar power, with China, India, the US, the Middle East, and Africa having the greatest potential. The Indian government has set out a "Solar Mission" that envisages that solar power will become cost competitive with coal by 2030.

3. Comparing the Costs

Energy is expensive. The costs fall on household budgets, all kinds of businesses, and on already constrained national budgets. So it's reasonable to ask which combination of these technologies can meet our needs at the lowest cost.

Unfortunately, that's a difficult question to answer in a truly objective way because there are so many uncertainties. The costs of renewables, especially PV, are coming down quite rapidly. We cannot be sure how long that trend will continue and how far it will go. Coal and gas prices are

---

[xxiii] International Energy Agency, *World Energy Outlook, Ibid.*

volatile. In the US in 2014 they were about half their peak levels in 2008.[xxiv] The cost of capital, the extent to which capacity is actually used, and the assumed cost of carbon emissions, are all important factors. As I have discussed, the situation in countries with different climates and natural resources may be very different. And then there is the question of whether the electricity will be available when needed. One cannot directly compare the costs of variable generation depending on the wind or sunshine with dispatchable generation from fossil plant which, in principle, is available at any time.

The IEA with the Nuclear Energy Agency (NEA) have published their estimates of the costs of electric generation from plant to be commissioned in 2020 based on national returns.[xxv] [Figure 6.3 below] They assume 85% utilisation of fossil plant, on the assumption that they will largely operate as baseload.

They also assume an environmental cost of $30 per tonne of carbon dioxide emissions. This is, inevitably, an arbitrary figure, in the absence of any international agreement. To put this carbon penalty into context, it adds about 15% to the costs of power from a modern gas plant in the US and 25% to a coal plant. On existing government policies, the IEA are assuming that the highest actual carbon prices in 2020 will be in Europe at around $22 per tonne of carbon dioxide. However, to meet the 2°C climate change target they would need to rise to around $140 in advanced nations by 2040. I have taken a discount rate of 7% p.a., the middle of the range offered by the study.

Coal and gas plant with CCS are not included in the study on the same basis because commercialisation is at such an early stage. The report estimates that, today, adding carbon capture to a coal fired plant increases its generating cost in the range of 30–70%. However, the costs of new plant built in the US in the 2030s could come down to $90 per MWh for coal and $65 for gas. That of course depends on the investment that is made in commercialising these technologies in the meanwhile.

What is one to conclude from these numbers? They demonstrate large cost variations from country-to-country. For instance, China and Korea have much the lowest nuclear costs, no doubt reflecting the benefit of

---

[xxiv] BP, *Statistical Review of World Energy*, 2015.

[xxv] International Energy Agency and Nuclear Energy Agency, *Projected Costs of Generating Electricity*, 2015 Edition.

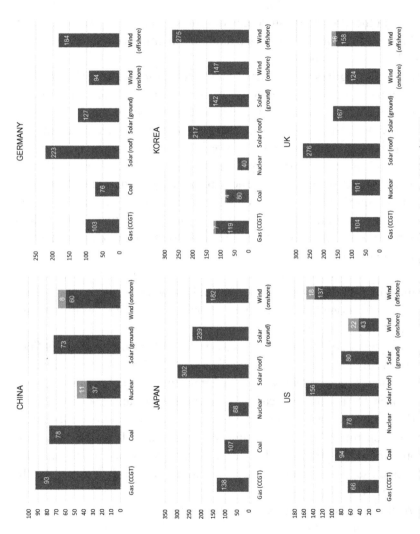

Figure 6.3. Projected Levelised Cost of Electricity, Commissioning in 2020; $/MWh.

Source: Projected Costs of Generating Electricity, IEA and NEA, 2015 Edition.

continuing construction programmes. Indeed the nuclear costs for the US and the UK, based on projections supplied by governments, seem rather optimistic bearing in mind the construction difficulties that new nuclear build has faced in the US and Europe. The costs of gas CCGT reflect low gas prices in the US, where shale gas has made such an impact, and relatively high ones in Japan which depends on gas imports. Wind and solar costs reflect local conditions.

Overall the figures demonstrate that, with a moderate allowance for the environmental cost of greenhouse gas emissions, the costs of onshore wind and solar power capacity installed in 2020 should be generally equal to or below those of coal and gas, the main source of electricity today. These technologies plainly have a bright future and, indeed, in 2016, nearly two thirds of additional capacity in all types of power plant was in renewables. The extent to which renewables may dominate the electricity scene in the future will depend on how we can adapt our energy systems to cope with variability, and the costs associated with that. This is discussed in much greater depth later on, in the sections on energy storage and the switch to renewables.

In spite of these promising cost comparisons, the fact remains that today most investment in renewables around the world is based on government subsidies of one kind or another. As one example, it's worth looking at the situation in the UK.

In the UK, in 2015,[xxvi] the wholesale price of electricity was about $60 per MWh. From 2014, the UK adopted a policy of supporting new investment in renewables capacity through competitive public auctions. In the 2015 auction for renewables capacity, onshore wind and PV came in at about the same level of $105 per MWh. That's a considerable mark-up for electricity consumers to pay, guaranteed for 20 years. It's higher than in most other countries because the UK is rather cloudy and good onshore wind sites close to the centres of demand in the Midlands and the South East are hard to find.

Part of the reason for subsidising renewables is to bring costs down for the future, and that does seem to be happening. Future subsidies are expected to be much less, if subsidies are needed at all. Nevertheless the

---

[xxvi] Carbon Brief, *UK Renewables Auction Pushes Down Costs*, February 2015.

prices agreed for the capacity in 2015 are locked in for 20 years. Its worth considering what value this represented in terms of greenhouse emissions saved. That of course depends on the carbon intensity of the fossil plant that the renewables will back out. If this is gas fired the subsidies work out at about $120 per tonne of carbon dioxide. If coal fired plant is displaced, the cost is about $55, since almost twice as much carbon is displaced by the same amount of power.

These are high prices, far above those in the European Trading Scheme, which peaked at around $34 per tonne of carbon dioxide and, in 2016, have traded well below $10. On the other hand, the IEA is projecting that if we are to meet the two degree target we will need carbon prices, in Europe and the US, of $100 by 2030 and $140 by 2040. On that basis, perhaps it isn't such a bad bargain. At the same time, in 2016, the UK government was considering a proposal to guarantee the price of electricity from a large new nuclear plant at $123 per MWh, much higher than onshore wind and solar, and even above the latest, 2017, auction prices for offshore wind, which were in the range of $75–$100 per MWh. Of course these prices are not truly comparable because nuclear is not intermittent. Hopefully, continuing auctions will continue to drive the price of renewables down.

Energy prices are a hot political topic in the UK in May 2017. As an election approaches, the Conservative government has promised that it will cap electricity prices if reelected. The UK's climate watchdog, Committee on Climate Change, estimates that about 9% of combined electricity and gas bills are attributable to climate policies, including renewables subsidies. That is $190–225 per household. By 2030, about a quarter of electricity bills will be due to climate policies. However, the Committee also points out that the savings that consumers have made from reduced energy purchases as a result of efficiency improvements more than counteract these increased costs.

The cost of energy matters a great deal to households who are "just about managing", to businesses that face stiff international competition, and, where subsidies are concerned, for strained government finances. However, as I have emphasised in this section, it is not always easy to judge the true costs of an energy option. Costs depend very much on location and they vary from time-to-time. Fossil fuel prices are volatile. Wind and solar costs have been coming down sharply and are still declining.

Costs are not independent of investment decisions because, as we have seen most spectacularly with solar energy, sustained investment can drive costs down. The costs of variable renewables are not directly comparable with those of baseload or dispatchable options, because they are providing different services. The costs of low-carbon energy are not comparable with those of fossil energy because of the externalities of climate change and local pollution.

Governments should be aiming to minimise the total system cost of providing reliable energy in a way that meets their environmental objectives, over an extended period. That may include investing in renewables that are not the cheapest source of energy today. It may also include investing in nuclear power or carbon storage that are expensive today but may create options that will be essential for the future. While government intervention is essential to bring forward new low-carbon technologies, the ultimate aim should be to create a regulatory framework, including low-carbon incentives, in which the market will determine the most efficient outcomes.

4. Energy Storage

Today, most electricity storage is in the form of hydro plant adapted so that water can be pumped back up from a low-level reservoir. This is a mature but costly technology. A wide range of other storage technologies are being researched. The cost of lithium-ion batteries for grid scale storage have fallen by more than three quarters between 2008 and 2013.[xxvii] They are still no cheaper, measured in terms of costs per storage capacity, than traditional lead acid batteries, but they can cycle more times and they are lighter and smaller — all critical functions for many applications, including vehicles. It seems certain that the use of storage will grow in many grids around the world.

It seems likely that storage will become significantly cheaper in the future. If so, it is possible that PV plus the necessary storage to convert its output to continuously available power will become a highly competitive source of electric power in years to come, especially in locations where

---

[xxvii] *Projected Costs of Generation of Electricity, Ibid.*

there is plenty of sun. We cannot be sure of that today, although we should certainly pursue research avenues that might achieve this.

By giving more flexibility on the timing of electricity flows, storage could ease constraints on electric grids and reduce the need for costly investment in new capacity. In countries where grids have a limited reach, storage creates new possibilities for "island" electric systems based largely on renewables.

However, very cheap storage could ultimately call in question the role of the grid. There is an irony here because, in the absence of cheap storage, enlarged and flexible grids are needed to accommodate high levels of variable renewables. A major role of the grid has always been to manage temporal variations in the levels of demand and supply by aggregating across large populations. With sufficiently cheap storage, individual domestic consumers or small groups with their own PV or other sources of generation may see no need to connect to the grid.

Indeed, in the most favourable locations, such as Hawaii, we are already on the cusp of this development.[xxviii] Hawaii has expensive electricity because the utilities largely burn fuel oil and because the different islands are not connected. And of course Hawaii has lots of sun. 12% of the utility customers already have rooftop solar, the highest proportion in the US. A US solar company, Solar City, are now offering a package of solar generation combined with Tesla's latest batteries designed to enable customers to cut themselves off from the grid entirely. Opinions differ on how great the uptake will be. The president of another solar company, ProVision Solar, has been quoted as being fairly sceptical. "People cry and moan about their utility bills, but compared to the time and effort that is needed to maintain an off grid system, it's night and day.[xxix] It's hard to convey unless you've actually lived off grid how different an experience it is". We shall see!

[xxviii] Mulkern, A. *SolarCity Plans to Sell Hawaii on off-grid Solar Package using Tesla Battery.* ClimateWire 6 May, 2015. Available at http://www.eenews.net/stories/1060018058. Accessed 5 October 2016.

[xxix] Climatewire, Electricity. *SolarCity Plan to Sell Hawaii on off-grid Solar Package using Tesla Battery*, 6 May 2015. Accessed at https://www.eenews.net/climatewire/stories/1060018058.

The potential for "island" power systems has already been mentioned. Renewables plus battery storage, sometimes with limited conventional power back-up, already represent an economically attractive source of power for many isolated regions, for instance in Africa or India, and for small islands.[xxx] The Isle of Eigg in the Scottish Inner Hebrides is an example. The power system includes solar PV, wind turbines, hydro-power, a bank of batteries, as well as a diesel generator. 90% of power comes from renewables. Residents are limited to a maximum of 5 kw, but this is seldom a constraint.

Besides cost, the environmental impact of battery storage also requires careful study. The energy input for manufacturing batteries is considerably higher than for mechanical forms of storage, such as hydro. Also, to varying degrees, batteries may contain materials that are in short supply or toxic. Disposal of lithium-ion batteries, one of the most promising technologies today, has to be carefully managed.

How expensive is electricity storage? In 2015 the investment bank Lazards estimated that, with expected cost declines over the following 5 years, the cost of storage for PV back-up might come down to a range of \$243–\$418 per MWh of electrical discharge.[xxxi] That seems rather expensive in comparison to other energy costs. But even at these prices, storage can play an economic role in filling in short-term gaps in local supply caused by transmission limitations or intermittent generation. The alternative might be to make costly investments in conventional plant that is very little used. That is why the market for storage is growing rapidly. Storage is also competitive in some off-grid situations. Nevertheless, it remains the case, in 2017, that, as a general rule, using renewable energy that had been stored in a battery for regular grid supply would be a costly option.

In August 2016, US Energy Secretary Moniz predicted that energy storage would lead to full decarbonisation of the US by 2050. "Energy storage technology is rapidly advancing and soon it may allow wind and solar to overcome their major flaw of intermittent generation. It could

---

[xxx] Grantham Institute for Climate Change and the Environment, *Electricity Storage for Climate Change and the Environment*, Grantham Briefing Paper No. 20, July 2016.
[xxxi] Lazard's Levelized Cost of Storage Analysis, Version 1.0.

even dispense with the need for nuclear and thermal power".[xxxii] He may be right, but this depends on achieving very big reductions in electricity storage costs.

Some of the cost reductions may come from large scale manufacture using already proven technologies. That is the aim of Tesla's "Gigafactory", opened in Nevada in 2016. However, improving the efficiency and reducing the costs of electric batteries is a top priority for scientific research around the world. There are many possible options and research teams around the world are working on this. For instance, scientists at Harvard are working on "organic flow" batteries which store chemical energy in external tanks using low-cost and non-toxic materials. These are bulky installations, but they could reduce the cost of power storage for electric grids.[xxxiii]

Scientists at MIT are working on a technology, the "lithium-air battery", that may have the potential to quadruple the storage capacity of lightweight batteries needed for mobile phones and vehicles. The research focuses on chemical storage of oxygen within the battery.[xxxiv]

These are just examples. There are many other lines of scientific research that could improve our ability to store energy and electric power. The best options for mobile appliances or vehicles, where weight and size are critical, may be quite different from those for stationary requirements, such as back-up for electric grids. However, experience teaches us that the passage from laboratory bench to mass adoption is not always smooth.

5. The Switch to Renewables

As already mentioned, renewable energy in the form of wind and PV is experiencing the most rapid rate of growth of all the electricity supply technologies. By 2016 they contributed 5% of global electricity supply, up from 4% in 2014. The IEA are projecting that, after improved energy

[xxxii] Retrieved from http://www.energylivenews.com/2016/08/18/us-to-decarbonise-by-2050-with-energy-storage/. Accessed 19 August 2016.
[xxxiii] US Department of Energy, *Developing Organic Flow Batteries for Energy Storage*, Project Impact Sheet, 24 February 2016.
[xxxiv] The Economist, *Lithium–air Batteries their Time has Come*, 6 August 2016.

efficiency, it is renewables that will have to play the most important role in our efforts to mitigate climate change.[xxxv] The scale of support for renewables, especially the variable renewables such as wind and PV, and the role that they are expected to play in the future, have become amongst the most controversial topics of energy policy.

In 2015, for the first time, more renewable generating capacity was added, world-wide, than fossil and nuclear capacity combined. At $288 billion, renewables also accounted for 70% of all investment in generating capacity.

These are staggering developments, although one has to remember that some renewables, because of their intermittency, deliver a considerably lower proportion of their stated capacity in actual generation. Also, with fossil plant one has to consider the investment in fuel supply as well as the investment in generating plant. Total investment in fossil energy, including coal, oil, and gas exploration and production still far exceeds that in renewables.

In the IEA's central projection, global investment in renewables totals $7.8 trillion over the period 2015–2040 and more than 60% of global power plant investment. Variable renewables contribute 40%. Even so, wind and solar only provide 19% of world electricity generation in 2040, well below that of coal, and only 6% of total energy.[xxxvi]

Today the back-up for variable renewables is fossil power and, in some cases nuclear energy. When the sun shines or the wind blows, renewable electricity supply displaces the alternatives, in most cases fossil generation, and undoubtedly reduces carbon emissions. However, renewable energy that depends on fossil power for back-up is not really zero carbon. Fossil plant providing back-up for renewables may include conventional coal or gas plant running on irregular schedules, low-cost "once through" gas turbines, or even diesel generators, and all of these are relatively high-carbon emitters. Maintaining fossil fuel or other back-up generating capacity alongside renewables is costly. The future of variable renewable power sources will depend partly, of course, on technological progress and cost reductions, but also on the development of power

---

[xxxv] International Energy Agency, *Energy Technology Perspectives*, 2016.
[xxxvi] *Ibid.*

systems as a whole and how they cope with high level of variable renewable power, a subject discussed further later in this chapter.

5.1. *Some Examples — Europe*

Germany

Germany has become the pioneer of renewable energy. In 2016, wind plus PV energy contributed 20% of electricity supply. However, renewables capacity was much greater than this. On 25 July 2015, when the sun was shining in the south, where most of the PV is located, and the wind was blowing in the north, where most of the wind generators are, renewable energy contributed 78% of electricity. At less favourable times wind and PV contribute very little. The system can cope because Germany still has large capacities of coal, lignite, and nuclear power. The government aims to phase out both nuclear and fossil fuelled plant but at the moment it is hard to see how this can be achieved without risks to energy security.

This great German experiment is leading to high energy prices (though not for industry, which is protected), strains on the grid, and big losses for major German utilities. According to *Fortune* magazine, Germany spent $26 billion on renewables in 2016, most of it in a surcharge on electricity bills.[xxxvii] However, according to the German energy think tank, Agora, in 2015 90% of the German public rated Germany's low-carbon energy transition, known as Energiewende, as important or very important. This share was unchanged since 2012. So it does not appear that the German public is losing faith. German experience demonstrates that, in the right circumstances, it is possible to win public support for radical energy change, even at considerable cost.

There is a paradox at the heart of German energy policy because, despite heavy investment in renewable energy, carbon dioxide emissions did not decline over the 5 years to 2015. This is because coal and lignite power production continued to represent more than 40% of German electricity generation. As renewables have increased, a growing share of German fossil electricity has been exported. The German authorities now have to find ways to reduce the operation of coal and lignite capacity

---

[xxxvii] Ball, J. *Germany's High Priced Energy Revolution, Fortune,* 14 March 2017.

while keeping it available as back-up when renewables are not available, a problem that can only get more difficult as Germany progressively closes its nuclear plants.

Another problem faced by German policymakers is that the predominance of renewable energy is in the north of the country while the heaviest industrial demand for power is in the south. The necessary strengthening of the grid is costly and there is strong resistance from the local public concerned about their impact on the landscape.[xxxviii]

In 2016, the German government decided on changes to its renewables policies, described by Minister Gabriel as a "paradigm shift".[xxxix] The "feed in tariffs" which have guaranteed subsidised prices for wind energy are to be replaced by a system of competitive government auctions. And the government will limit the share of renewables in electricity generation to 45% in 2025, which implies some moderation of the investment rate. There is no change in the basic objectives of Energiewende, and Germany still aims to have 80% renewable electricity by 2050. But there is certainly concern in the renewables industry as to what the future holds as the government strives to control the cost to consumers of its renewables policy.

UK

In the UK, wind and PV contributed 15% to electric power in 2016. However, the growth of renewables, which run ahead of fossil generators because of their low variable costs and guaranteed output prices, has discouraged investment in other power options. There is growing concern about the security of supply at peak winter demand and the government has instituted capacity payments in the form of subsidies to promote investment in new fossil energy or keep existing plant available. Recent rounds have mainly financed new diesel generators. This is not necessarily a problem if these are only used for a few hours per year. But it's a reversal of environmental policy if they come into more regular use.

[xxxviii] New York Times, *Germany's Clean-Energy Plan Faces Resistance to Power Lines*, 5 February 2014.
[xxxix] International Energy Agency, *Projected Costs* etc, *Ibid.*

The UK, like Germany is moving from a system of fixed "feed in tariffs" to government auctions in a bid to manage the cost of subsidies better. The UK has also, in 2016, abolished its Department of Energy and Climate Change (DECC) and merged energy policy into a new Department of Business Energy and Industrial Strategy (BEIS), a similar structure to that of Germany. This is not supposed to signal any weakening in environmental commitments, but some environmentalists see it as a worrying development.

Denmark

Denmark pioneered the development of wind energy and its wind industry is a big contributor to Danish exports. In Denmark the share of wind energy in electricity generation has achieved the impressive level of 43%, the highest in the world. This is possible partly because Denmark's power system is connected to the grids of Norway, Sweden, and Finland, through the Nordic Power Pool, and these countries have large capacities of hydro, nuclear, and fossil fuel. Denmark has set itself the objective of being free from fossil fuels by 2050. Denmark, like Germany, already has higher domestic electricity prices than the rest of Europe, and this is due in part to the cost of renewables support.

In May 2016, amid growing concerns about the cost of wind subsidies, the newly elected liberal alliance Danish government announced that it was to scrap the Public Service Obligation, the levy on electricity bills that is used to finance wind energy, describing it as "expensive and ineffective". Whether this is so may be matter for debate, but it seems clear that the Obligation had become unpopular. The government has not scrapped the 2050 fossil fuel free target and will look for other ways to achieve it, possibly with a somewhat reduced reliance on wind. It remains to be seen how this will impact on Denmark's green ambitions, but there are certainly signs of tension. The government is calling its new approach "green realism".

Spain

In Spain, wind energy alone accounts for 19% of electricity supply. Spain has a highly diversified power sector, with major contributions from coal, natural gas, and nuclear power, as well as from renewables, including hydro. However, Spanish support for renewables has been sharply cut back in the last few years as the country has faced an economic and financial crisis. Spain has the most intense sunshine in Europe and until 2011 was one of the world's biggest markets for PV. However in that year Spain not only reduced support for new projects but also retrospectively cut support promised to existing projects, a move that severely undermined business confidence. PV investment fell dramatically as a result.

More recently, in 2015, Spain introduced a levy on domestic PV generation, intended to contribute to electricity network costs. There is some logic in asking PV generators who remain connected to the grid to pay their fair share of system costs, although this has to be weighed against the desirability of encouraging PV deployment. The move has been widely criticised as a "tax on sun" and, as of 2016, it seems likely that the levy will be withdrawn. Overall, current Spanish policy is for continuing, but much slower, growth in renewables combined with greater reliance on gas. Spain is also planning additional grid connections to France, which should enhance the flexibility of both national systems.

These examples illustrate the rapid progress, but also the growing pains, of variable renewable energy in Europe, where their market shares are highest. The positive story is that as the renewables technologies mature the costs are coming down and they are getting closer to the day when they can stand on their own feet. On the other hand there are also signs of renewables subsidy fatigue and, as the share of variable renewables increases, problems of balancing the system are emerging. This may explain why the rate of renewables investment slowed, somewhat, in 2016.

Summary

Renewables are set to play a much bigger role in the energy landscape of the future. But for renewables to become dominant will require a revolution in the way that electricity systems work. We shall return to this in Section 9. The jury is still out on the question of how successful the most progressive European countries will be in managing continuing increases in renewables penetration.

Europe, of course, is not alone in investing in renewables. Indeed, as deployment in Europe has slowed, the market in developing countries has been growing rapidly and the biggest markets for renewables are now in Asia. However, these countries are still well behind the most advanced European countries in terms of renewables' share of electricity supply.

5.2. *Some Examples — Asia*

South East Asia

South East Asia contains some of the world's most dynamic economies. Its GDP has more than doubled between 2000 and 2014 and electricity demand nearly tripled. Governments are primarily concerned to ensure a rapidly growing electricity supply at affordable cost to improve social welfare and support industrial development. The region has ample supplies of coal and in the medium-term coal is expected to remain amongst the lowest cost and most accessible form of energy. The region also has gas but, at current prices, it makes economic sense to export gas and use coal domestically. As a result, heavy investments in new coal capacity are planned.

The power mix today is 44% gas, 32% coal, and 6% oil, but coal is expected to overtake gas in the next few years.[xl] The region plans to add some 88 GW of coal capacity between 2015 and 2030, which is more than the total generating capacity of the UK. The adoption of efficient supercritical and ultra-supercritical coal technology will be important to moderate the growth of carbon emissions. Coal power stations have lives of 40–60 years and this means that, to achieve very low levels of carbon emissions in the future it is likely that some of this plant will either have to be retired early or to have CCS fitted retrospectively. Only the most efficient plant will ever be suitable for CCS and this is another reason why it is important to employ the latest technologies.

The countries of South East Asia are open to strategies that would rely more heavily on renewables, including geothermal and hydropower for

---

[xl] International Energy Agency, Insights Series. *Reducing Emissions from Fossil Fired Generation, Indonesia, Malaysia, Vietnam,* 2016.

which the region is well endowed. There are also growing signs that the rapidly falling cost of solar PV may have the potential to disrupt today's coal based plans. But the governments concerned will have to be convinced that this is consistent with their primary objectives for economic and social development and this will require the technical and financial support of the developed world.

India

There are some similarities with the situation in India.[xli] India has the world's largest population of very poor people and, in 2013, more than 230 million people without access to electricity. There are strong humanitarian reasons for helping India to turn this situation around. Understandably, The government's overriding objective is economic and social development through rapid economic growth. India already has a severe power shortage and needs to increase its generating capacity rapidly by the most economic means. India also has large reserves of low-cost coal. The government has announced an ambitious plan to more than double production, to 1.5 billion tonnes p.a. by 2020.

India also has excellent renewable resources, including hydro, wind, and solar and has announced a national mission to promote solar power. But on present plans, coal will still dominate India's power sector at least to 2040, by which time India may have become the world's second largest carbon dioxide emitter, exceeded only by China. The development of India's electricity distribution system has not matched the growth of India's economy and there are chronic supply difficulties. Many of the problems are administrative and political rather than technical. These include political interference in power markets, tariffs held below cost, theft of power, and sanctioned non-payment by government agencies. There are also local difficulties and delays in securing sites for new power stations. It makes sense, consistent with this objective, to support India's efforts to deploy renewables and to implement the most efficient coal technologies, to prepare for the possibility of CCS in the future.

[xli] International Energy Agency, *World Energy Outlook*, 2015.

China

In 2014, China was still 73% dependent on coal for its electric power, but it now also has the world's largest renewables sector. More than half of the new onshore wind capacity added in 2015 was in China. In 2016 there was a major rebalancing towards solar. China added an amazing 35 plus GW of solar power, about half the world total, whereas additions to wind power shrank to about half that level. China suffers from too much generating capacity and its management of the competition between renewables and fossil generation makes an interesting contrast with Germany.

China has a wide ranging problem, which the government is trying to address, of excess investment intended to boost economic growth. As the economy rebalances to less energy-intensive sectors, the growth of electricity demand is slowing. The government has set ambitious targets for investment in renewables but, so far, has not been able to halt the continued investment in fossil energy.[xlii] At least 50 GW of new coal plant were built in 2015.

In Germany and most other developed economies, renewables run ahead of fossil plant because of their lower marginal costs. In China, the system is administered. There is serious "curtailment" of renewables power, in which about 30% of available capacity is not used. This is partly because of limitations in the capacity of the grid. But it mainly stems from the administration of power system dispatch. All power stations are given quotas as to how much they can run but the utilities have a certain amount of flexibility to choose different power sources at the margin. Renewable power tends to be priced higher than coal and the big state owned utilities are sympathetic to the coal industry so this system tends to favour coal over renewables. Chinese authorities are embarking on a major reform of the system to make it more market driven, but this will inevitably take some time.

As discussed in Chapter 4, China is undergoing an ambitious programme of power industry reform, intended to make the whole system more market oriented and efficient. It is to be hoped that this will lead to

---

[xlii] Edward Wong, *Glut of Coal-Fired Plans Casts Doubts on China's Energy Priorities, The New York Times*, November 2011.

much fuller use of existing renewables capacity. Until then the problems of absorbing existing renewables capacity are bound to cast a shadow over the ambitious new investment plans.

5.3. *The Progress of Renewables*

2015 was the first year in which renewable energy excluding large hydro made up the majority of new power generating capacity (56%).[xliii] It was also the first year in which investment in renewables in developing countries outstripped the investment in developed countries. However, investment in developed countries was well below its peak, which was in 2011. Investment in Europe was less than half its peak levels, with big falls in Germany, Spain, and France. Investment in the US held up, no doubt because of fears, which turned out to be unfounded, that key tax credits were about to be withdrawn. Japan continues to be a massive investor in small-scale PV.

Of the developing countries, China, India, and Brazil are the big three. The growth of renewables in China has been spectacular and in 2015 China alone accounted for more than a third of global renewables investment. Besides India and Brazil, South Africa, Mexico, Chile, Morocco, Uruguay, and Thailand, all invested $1 billion or more, and there was widespread investment, on a smaller scale, in other developing countries. Overall, there was a healthy rate of growth, of 5% over 2014, although the level in 2015 was not much higher than in 2011, at the height of the German and Italian rooftop solar booms. The big news in 2016 was the spectacular growth of solar PV.

The growth of renewables in developing countries is excellent news and a good sign for the future. However, in volume-terms it remains very dependent on China. As explained above, China needs to address its "curtailment" problem to get full value from its renewables investments. The decline in investment in developed countries, especially in Europe, suggests that the path towards very high level of renewables penetration may not be smooth.

[xliii] *Global Trends in Renewable Energy Investment 2016*, Report by the Frankfurt School-UNEP Collaborating Centre for Climate and Sustainable Energy, in collaboration with Bloomberg New Energy Finance, for UNEP.

6. Looking Beyond Electricity

In 2016 less than 40% of world energy was used to generate electric power. The other 60% was used in buildings and in industry, mainly to generate heat, and for transport. Similarly, only just over 40% of world carbon dioxide emissions were attributable to electricity. Figures 6.4 and 6.5 show the composition of total world energy supply in 2016 and how the IEA project that this might need to change by 2040 if we are to meet the target of limiting climate change to 2°C.

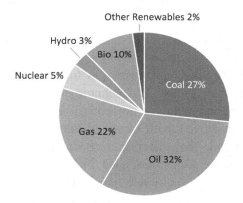

Figure 6.4. World Energy Demand in 2016.
Source: IEA, World Energy Outlook 2017.

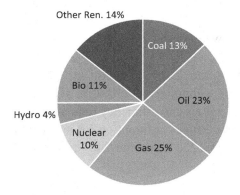

Figure 6.5. World Energy Demand in 2040, on 2 Degree Course.
Source: IEA, World Energy Outlook 2017.

We now look, in turn, at the main sectors of energy demand that are not presently dominated by electricity.

7. Transport

7.1. *Vehicles*

In 2016, electricity contributed only about 1% of world energy for transport, and this is mainly for railways. 90% is oil. Oil contributed 35% of global carbon emissions in 2016, and most of this was used for transport. The internal combustion piston engine, which in some ways appears today a rather primitive concept, has proved extraordinarily resilient. At different times it has seen off challenges from electric power, steam power, rotary combustion engines, and micro gas turbines. It still has a lot of potential for further refinement. The technologies that will make the most difference to oil demand and to carbon emissions in the next decade or so will be those that improve the efficiency of internal combustion powered vehicles. These include fuel injection, turbocharging, and variable valve timing. Ironically, these are technologies often associated with high powered and high petrol consumption vehicles. That is because, in the past, efficiency improvements have tended to be used to make heavier vehicles accelerate faster. Now, they are needed to reduce fuel consumption and emissions while maintaining performance. There is also a lot of potential to make vehicles, including engines, lighter. Unfortunately, however, the current trend is towards SUVs, whose market share was rising rapidly in 2016 and which now represent more than a quarter of global new car sales.[xliv]

Notwithstanding the 2015 scandal, in which Volkswagen were accused of installing technology to cheat on vehicle emissions tests, the achievements of emissions and efficiency regulation are considerable. The EU target fleet average fuel consumption for new cars in 2015 represents an 18% reduction from 2007, and the 2021 target a 40% reduction.[xlv] In

---

[xliv] JATO Dynamics,Global New Car Sales Q1 2016.
[xlv] The International Council on Clean Transportation, *Global Passenger Vehicle Standards*, August 2015.

the US, under the latest regulations, fuel efficiency is set to approximately double by 2025. Partly as a result of these standards, oil consumption is now in steady decline in the developed OECD countries. In the developing countries, where vehicle ownership is rising fast, oil demand is rising rapidly. However major developing countries, such as China, are also setting increasingly tough emissions standards.

Today the internal combustion engine is under threat not only because of climate change, but also because of the severe environmental and health effects of urban air pollution. There is considerable potential to mitigate local pollution effects, especially in the developing world, through higher fuel standards. For instance fuel standards in China are considerably lower than in the West. But this does not address the climate impact.

Switching to biofuels reduces the climate impact of the whole plant to fuel cycle but, as discussed in the section on biofuels above, the availability of plant material from environmentally acceptable sources is limited. Many experts believe that, in the low-carbon world of the future, biofuels will have to be reserved for the uses where alternative low-carbon options are most difficult to find, that is to say, heavy vehicles and aeroplanes.

What other alternative options are available? One option that has been widely canvassed, and is already being tried in some parts of the world is to use hydrogen as an energy vector. Hydrogen can be used to drive vehicles either in an internal combustion engine or by generating electricity in a fuel cell. In either case the exhaust product is water. Hydrogen can be generated through electrolysis of water using low-carbon energy.

There are hydrogen fuel cell cars on the roads in Japan today. The Japanese government has set a target of having 40,000 hydrogen fuel cell vehicles on the roads by 2020.[xlvi] The main problem is the huge infrastructure that would be required to deliver hydrogen to service stations. Someone has to place a very big bet before hydrogen vehicles can really take off. However, hydrogen may be the ultimate answer if it proves difficult to overcome the range limitations of electric cars.

[xlvi] Scientific America Climatewire, 3 May 2016, *Japan Bets on a Hydrogen Fuelled Future*. Available at https://www.scientificamerican.com/article/japan-bets-on-a-hydrogen-fueled-future/. Accessed 6 October 2016.

Electricity has the big advantage that conversion can proceed step-by-step. We already have an electric network. And electric hybrids and plug in hybrids can pave the way for full blown electric vehicles. Eventually major investments will be needed in the electric grid and in charging stations, but that can proceed in parallel with the development of the market for electric vehicles and the testing of public opinion. The fundamental problem with electric vehicles, which has slowed their introduction, is that the power density of electric batteries is much less than that of petrol. This implies greater weight and reduced range. Also, batteries are expensive. The costs are coming down with mass production but there has been no fundamental breakthrough in recent years.

Today almost all major motor manufacturers have introduced or are introducing electric models and, for instance, Tesla Motors are blazing a trail with exciting sports models. Sales of electric vehicles are rising fast, but from a very low base. The market share of electric vehicles in 2015 was in the range of 1–2% in the US, major European countries, and China.[xlvii] The share is higher in California and much higher in Norway, where almost a quarter of the market for new cars is plug-in hybrid electric. Norway is one of the richest countries, per person on the world, ironically, in this context, partly because of its substantial oil and gas exports, and Norway also has abundant low-carbon hydroelectricity.

The EU has set itself the target of reducing greenhouse gas emissions by 80% from 1990 levels, by 2050. Other advanced countries will need to set similarly tough targets if we are to meet the 2°C climate change target. 45% of the EU's carbon emissions are from transport, mainly passenger vehicles and trucks. Since there are some other sectors of the economy that will be very difficult to decarbonise, such as cement and steel, and trucks are more difficult to decarbonise than passenger vehicles, this implies a high degree of decarbonisation of passenger vehicles by 2050.

There are some alternatives to electrification, notably hydrogen which I have already discussed. Advanced biorefineries could increase the availability of vegetable based vehicle fuels, but they will always be in limited

---

[xlvii]Yang, Z. *2015 Global Electric Vehicle Trends: Which Markets are up (the most)*. The International Council on Clean Transportation, 18 May 2016.

supply. Most probably this means that to meet climate targets we need an almost total switch to electric vehicles combined with the necessary investment to deliver low-carbon electricity. Since the average age of vehicles in the EU is 10 years, new internal combustion vehicles would need to have been phased out by about 2035 for virtually the whole fleet to be low-carbon by 2050.

Is that possible? It's a social question as much as a technical question. It isn't impossible, bearing in mind that, according to their top manager, Matthias Muller, Volkswagen are gearing up to produce 2–3 million pure electric cars by 2025,[xlviii] which would be about a quarter of their total production today. Governments will need to provide support through subsidies for early adopters and investment in charging infrastructure. To a certain degree this is happening already. As market share increases there will also need to be substantial investment in the electric grid.

There would be other benefits from switching to electric vehicles. The contribution that vehicles make to urban air pollution is now causing such concern that, as soon as it seems credible socially and technically, the governments of major cities are likely to drive this change forward. Already London offers a rebate for hybrid and electric vehicles against its "congestion charge" for driving into the city centre. Governments and cities in Europe, the US, and China offer a wide range of tax and other incentives for electric and hybrid car ownership. National governments can also be expected to give a lead and already, in 2017, the UK and French governments have said from 2040 all new cars sold will have to be at least partly electric. Electric vehicles can also contribute to the flexibility of the grid, and its ability to manage variable generation. We shall come back to that point later in this chapter.

Public attitudes, in other words consumer choice, are the big unknown. Possibly electric cars will come to be the vehicles of choice. Depending on technological progress, electric vehicles may remain more expensive than conventional vehicles and some limitations in terms of range and charging times may also continue. Governments and environmental

---

[xlviii] Electrek, 16 June, 2016. Available at https://electrek.co/2016/06/16/vw-2-3-million-all-electric-cars-2025. Accessed 29 August 2016.

campaigners may need to convince the public that it makes sense to accept these limitations and pay a little more for their vehicles in the interests of the local and global environment. The price of electric vehicles is already well within the range that people pay for luxury cars!

The outlook is further complicated by the two other revolutions that may be in prospect for personal mobility, that is to say, driverless cars and the service approach to transport. In decades to come it may be that few people will own cars and, instead, we will subscribe to a service that provides self-driving vehicles on request. It isn't obvious how that would affect the choice of drive technology. Perhaps if vehicles become less of a statement of personal identity, and more of a commodity, governments and environmentalists will encounter less resistance to new technology that makes more environmental sense.

Much publicity has been given to the fact that the reserves of energy companies, including the oil industry, now far exceed the amount of fuel that can be consumed consistent with the 2°C global warming target. The way to address this problem is by reducing the demand for oil by making vehicles more efficient and switching to electric vehicles, as well as by changing life styles and using more public transport. So long as we all still depend on oil, there is not much value, or indeed logic, in criticising the oil companies for producing it.

7.2. The Other Uses of Oil: Shipping and Aviation

Trucks

One of the reasons why the oil industry is less concerned than some might assume about the advent of electric cars is that passenger vehicles account for only about a quarter of world oil demand.[xlix] Another 14% of oil demand is for power generation and industrial boilers, and most of this could eventually be substituted with renewables or gas. But substituting the other main uses of oil will be more difficult. Trucks account for 18% of oil demand and this share is gradually increasing. Oil still contains a lot more energy per weight than electric batteries and this makes it difficult to switch to electric power for heavy duty vehicles.

[xlix] International Energy Agency, *Energy Technology Perspectives*, 2016.

However, this section is mainly about boats and planes. They provide interesting examples of how the international community is trying to manage carbon emissions that cannot be attributed to one country. In 2015, aviation contributed 6% of world oil demand and shipping 5%. Emissions in both sectors are growing fairly rapidly.

Air Travel

In 2015 nearly 3.6 billion passengers were carried by the world's airlines, equivalent to about half the world's population. This staggering statistic underlines the fact that air travel, once the preserve of the rich, is now becoming accessible to a much larger share of the world population.

World passenger miles increased by an average of 5.5% p.a. in the 10 years to 2016, rising to 6.3% in 2015–2016. Year-on-year growth in the Asia Pacific region was 8.3%, in the Middle East 11.8% and in Africa 7.4%. It would be a brave government that took measures to interrupt this great expansion of opportunity.

Advanced biofuels or hydrogen are both possible low-carbon alternatives to mineral oil-based aviation fuel for the future. But they are longer, or perhaps medium-term, options. Today, they are far from being commercially viable on the necessary scale.

About a third of the operating costs of airlines are made up of fuel.[1] So there has always been a strong incentive for the airlines and their manufacturers to improve the efficiency of aeroengines, reduce the weight of aircraft, and increase seat occupancy. Today's aircraft may look fairly similar to the first jets of the 1960s, but they are about 80% more fuel efficient. As a result of these continuing advances, the growth in fuel use and carbon emissions, projected by the IEA at about 2% p.a. is much lower than the growth in passenger miles. The IEA expects that the oil used by the industry will have increased by more than 50% by 2040, by which time it will represent 9% of total oil demand.

Aviation may not account for a particularly large share of total carbon emissions, but any significant share that is moving in the wrong direction represents a challenge for the ultimate objective of a very low-carbon economy.

---

[1]Rock, N, *et al. First Ever Global Regime for Aviation Emissions: ICAO Adopts Global Market Based Measure to Combat Aircraft CO_2*, Reed Smith Client Alert, 10 October 2016.

From 2012, the EU included aviation in its Emissions Trading Scheme.[li] The airlines were given free allowances to cover most of their operations but were required to purchase additional allowances at the margin. Originally this was intended to cover all flights to and from Europe. But such was the outcry from almost all other major countries, including the US, China, and India, that the EU agreed to "stop the clock" on external flights, pending discussions on a global aviation scheme. These discussion were held at the UN's aviation agency, the International Civil Aviation Organisation (ICAO).

In 2016 ICAO agreed the framework for a global scheme for international flights, intended to come into effect in 2021.[lii] However, it will not be mandatory until 2027. Airlines will be expected to purchase carbon credits to cover the excess of the aviation industry's emissions above the 2020 level. The definition of eligible credits is till for discussion but probably the scheme will link to the UN's existing carbon trading mechanisms. There is a lot of work still to be done. But probably, for the scheme to win the necessary agreement from all major players, its impact will have to be fairly modest.

Aviation is a difficult case for carbon incentives. Because the industry is already highly motivated to minimise its fuel use, the scope for further efficiency gains is fairly limited, at least until revolutionary new technology becomes available. If the scheme simply raises the cost of flights and reduces carbon emissions mainly by preventing poorer people from flying it will not be very popular.

Shipping

World shipping is expected to grow more or less in line with the increase in world trade, since about 80% of international global trade in goods goes by ship. The IEA are projecting that shipping will grow by 3.6% p.a. to 2040. Most ships are powered by heavy fuel oil or diesel. However the

---

[li] IATA, *European Commission Proposes Intra-European Flights Should Remain Covered by EU ETS Pending ICAO Process*, 3 February 2013.
[lii] International Civil Aviation Organisation, Carbon offsetting and Reduction Scheme for International Aviation (CORSIA), 2016.

total fuel consumption, and carbon emissions, from ships are only expected to grow at 1.9% p.a. as ship sizes increase and engine performance and hull designs improve. By 2040, the share of shipping in oil demand is expected to have increased slightly from 5% to 6%.

Like aviation, international shipping is not directly included in the Paris climate agreement. However the UN's shipping organisation, the International Maritime Organisation (IMO) has adopted internationally binding efficiency standards, known as the Energy Efficiency Design Index.[liii] This came into force in 2013, and mandates a 10% efficiency improvement in new ship design from 2015, 20% from 2020, and 30% from 2025. The IMO is also working towards the regulation of sulphur emissions, which is the other major environmental concern that arises from the propulsion of ships.

There is no immediate prospect of finding alternative hydrocarbons for shipping. The use of biofuels or LNG can make some contribution to reducing emissions, as can wind assistance, through kites. Since about a third of international seaborne trade is in energy products, especially oil, reductions in fossil fuel demand should, eventually, also lead to some reduction in shipping.

8. Industry

Industry accounts for 37% of world energy use and 20% of direct carbon dioxide emissions.[liv] The biggest consumers are chemicals and petrochemicals, iron and steel, cement, and paper. There is enormous potential for energy savings through switching from coal to gas, more efficient plant and processes, and product design and recycling. The technologies that can make industrial processes less carbon intensive are, of course, highly diverse. But important common elements include the efficient design and "right sizing" of pumps and valves, heating and power circuits, waste recovery, product recycling, and day-to-day economic management. As described in Chapter 4, the drives to improve industrial energy

---

[liii] International Maritime Organisation, *Low Carbon Shipping and Air Pollution Control*, IMO Website, 2017.
[liv] International Energy Agency, *Energy Technology Perspectives*, 2014, p. 47.

efficiency in countries such as India and China are among the most important carbon reducing efforts in the world.

However, electrification is a difficult or impossible option for much of heavy industry and this means that, with currently available technology, CCS, capturing and storing underground the emissions from the burning of fossil fuels, appears essential to get the emissions from industry down to really low levels.

9. Buildings and Appliances

About half of global electricity is used in buildings, the largest part in homes. There have been great improvements in the efficiency of energy appliances, especially lighting, and this is helping to contain the growth in demand. However, at the same time, there is a continuing proliferation of the range of electrical devices and, fortunately, in the developing world the number of people with access to modern homes and appliances is growing rapidly. So the decarbonisation of electricity is the most important step for reducing carbon emissions attributable to buildings, and, conversely, improving the efficiency of buildings and their appliances is amongst the most important measures for moderating the growth in low-carbon electricity that will be required.

In cold climates, including most of the developed OECD countries, 60% of the energy used in buildings is for space and water heating. Most of this is provided by fossil fuels, especially gas. For instance, in the UK gas from heating homes contributes nearly 20% of total carbon emissions from energy. For the UK, and other developed nations in a similar situation, reducing carbon emissions to very low levels will require that alternative options are used, most probably electricity with heat pumps.

Heat pumps operate like reverse refrigerators, concentrating the heat in the air or in the ground. They can deliver between two and six times as much heat energy as the power needed to drive them. Nevertheless, the heat that they provide is fairly expensive, and because they do not generate high temperatures, they are really only suitable for homes that are already well insulated.

Ultra efficient buildings with very low heating demand in temperate climates are perfectly feasible. They have high levels of insulation throughout, including doors and windows, well oriented windows, and materials to absorb heat, and they are either largely sealed against air coming in from outside or they have heat exchangers so that this air is prewarmed to room temperature.

Getting newly built houses up to these standards is a challenge, but for a country like the UK, with an ageing housing stock and which only about 1% of additions each year, the main problem is raising the standards of existing housing. Programmes to insulate lofts, install doubler glazing, and inject insulation into cavity walls have all been reasonably successful. But the task of improving the insulation of the basic building envelope, has not really been tackled.

Housing in the developing world, measured by floor area is set to increase by more than 80% by 2030 and by a further 45% by 2050. Most of this huge building programme will be associated with rapid urbanisation. For buildings in cool climates the quality of insulation of these buildings will be crucial. However, many developing nations have hot, humid, climates, and here the need is for buildings that minimise the need for air conditioning. Reflective roofs and walls, exterior shades, and window coatings and films can all reduce the need for cooling.[lv]

10. A Flexible System for the Future

How will all these technologies combine to deliver the clean, affordable, and convenient energy that we will need in the future? This is the biggest question.

Many experts hope and expect that electricity from renewables will become the dominant energy source. In addition to powering appliances, as it does today, electricity will need to provide the energy for transport and heating as well as large parts of industry. Enlarged and highly sophisticated grids will balance demand and supply, drawing from a wide range of sources across broad regions. The variability of wind and solar power will be accommodated in part by other renewable sources, such as bio or hydropower. Batteries, including the batteries of electric vehicles, or other storage systems, will even out supply. Fossil fuelled capacity may still be needed as the ultimate back-up but it will run only for very short emergency periods, making the emissions negligible.

[lv] IEA Technology Roadmap, *Energy Efficient Building Envelopes*, 2013.

In this scenario, the demand side will also contribute flexibility. In other words, domestic and industrial electricity users will adapt their demand patterns to the availability of renewable energy. The hope is that the way people and businesses buy electricity will change to a "service" basis. Instead of buying electricity itself we will buy the services that it provides; warmth, functioning of appliances, perhaps mobility. Perhaps other services will be included in the package. The companies that supply this will meet our needs in the most efficient, least expensive way, using all the resources of information technology and intelligent appliances. Rooms do not need to be heated when we are out, washing machines do not need to run when electricity is expensive, car batteries can be charged at night and perhaps discharged when their power is needed. In a world where variable renewable energy is a large part of electricity supply there will be big economic benefits from managing demand in this way. Smart companies, like Google or Amazon, experienced in applying IT to consumer needs, might play a role in finding ways to make this work.

A change of this kind seems likely, but the extent and pace of change are uncertain. Some people point to the revolution in communications, and the adoption of mobile phones, as an example of rapid change in consumer habits based on advanced technology. But unlike the communications revolution, there is no benefit to consumers in the adoption of "smart energy". We already have electric power that is reliable and fully responsive. Costs will probably rise. A service contract will have to specify the customer's needs in some way and it will be hard to avoid an element of restriction. There is a long history of energy service companies supplying industry and experience has been mixed because of the difficulty of defining energy needs. Possibly some persuasion will be needed from governments and environmental groups pointing out the importance of service supply for the adoption of green energy. Efforts to persuade households to recycle their waste, also for environmental reasons, have been very successful so perhaps that is a positive sign.

In their Energy Revolution report,[lvi] Greenpeace, together with international solar and wind bodies, offer an Energy Revolution scenario in

---

[lvi] *Energy Revolution, A Sustained World Energy Outlook 2015*, Greenpeace, Global Wind Energy Council, Solar Power Europe.

which 83% of all energy is supplied by renewables in 2050. This requires dramatic improvements in energy efficiency so that total energy demand in 2050 is only slightly more than half the level in their Reference Case.[lvii] Improvements in energy efficiency play the largest part in emissions reduction. Wind generating capacity is four and a half times that in the Reference Case, and solar PV is eight times. 55% of electricity comes from variable renewables, so smart grids, energy storage, and demand side management have to provide the necessary flexibility to keep the lights on. Greenpeace recognises that:

> Further research and development of innovative technologies is a pre-condition of the Energy Revolution and the required fundamental trans-formation of the energy systems. Thus, the scenarios assume further improvements of performance and costs of plants for electricity and heat generation from renewables, new vehicle concepts, and in an implicit way also for other infrastructures such as electricity and heating grids, charging infrastructure, synthetic fuel generation plants or storages and other load balancing options.[lviii]

We should certainly find cost effective ways to increase the flexibility of grids and make the best use of renewable energy. But we are a long way from achieving this kind of reliance on renewables today. Wind capacity factors are in the range of 20–49% and it is possible, for instance in the UK, for cold weather with little wind to persist for several days. In some countries, including northern Europe, peak electricity demand is in the hours of darkness. So there is a hill to climb. A great deal depends on whether the costs of wind, and PV, continue to decline, and on achieving very large reductions in the cost of energy storage.

It is not at all certain that renewables plus flexibility plus storage will provide the complete answer to low-carbon energy supply, at least for a long time to come. Most probably we will continue to need other forms of more reliable electricity. Ideally, this should come from low-carbon sources such as nuclear power or fossil generation with CCS. Almost certainly CCS

[lvii] The Greenpeace Reference Case is based on the Current Policies case in the IEA's 2014 World Energy Outlook.

[lviii] Greenpeace *et al.* 2014, *Ibid.*

will be needed for some fossil fuel dependent industries. In practice, gas generation will be important during a transition period of many decades.

I have described how a variety of energy supply technologies could be coordinated with demand flexibility through a sophisticated power grid. This is what one might call the "all singing, all dancing" option. But there is an alternative, and to some extent competing, option. If storage becomes sufficiently cheap it may make sense for some electricity consumers, especially domestic or commercial consumers in hot countries, to cut themselves off from the grid altogether and rely on their own PV plus stored power. This could hollow out of the economic basis for the grid. That is not a problem if the grid can be dispensed with altogether, but if the grid is still needed for some domestic consumers, for large scale renewable energy such as offshore wind arrays, and for heavier industrial use then government regulating agencies are going to be faced with a dilemma. This isn't an immediate problem. Today utilities in sunny regions such as California or Spain are more concerned with the concessions being granted to PV generators who are connected to the grid. But the self-sufficient PV systems that are being offered in Hawaii will no doubt become widely available.

11. Developing Countries

Many of the technologies for developing countries are not in principle very different from those needed in the developed world. The key requirement is sufficient energy to sustain economic growth. In many poorer countries inadequate supply leads to frequent blackouts, with consequent disruption. One result is that wealthier individuals and businesses maintain their own back-up supply, often in the form of diesel engines, one of the least efficient, most expensive, and most polluting options.

For communities relying on traditional biofuels, such as wood or animal dung the first steps may be to provide more efficient and sustainable use of biomaterials, and especially clean cooking stoves which, as explained in Chapter 4 are vital to improve health. Switching to LPG can also be part of this process. Because the needs of traditional communities are relatively modest, the impact, from the perspective of climate change, need not be significant.[lix]

[lix] International Energy Agency, *World Energy Outlook*, 2016, p. 419.

Access to electricity can be provided by local mini-grids or extension of the main grid. In their 2014 report on African energy, the IEA studied the particular needs of Nigeria and Ethiopia.[lx] In Nigeria, which has a high population density and a relatively developed national grid, grid-extension is the most cost-effective route to increasing energy access. In Ethiopia, where the population density is considerably lower, a significant proportion of the population still lives in areas where grid extension is the best option, but off grid solutions have a bigger part to play. Typically the cost of electricity from grid extensions is well below that from mini-grids. Most of the power for mini-grids is provided by diesel generators and solar PV, whereas PV has the leading role in multiple off-grid systems each supplying a small amount of electricity.

There is a great attraction to the idea of meeting the needs of remote communities in Africa or Asia with renewable energy. Indeed, the US development agency's "Powering Agriculture" scheme is restricted to renewable sources. However, as Sarah Best says in her discussion paper for the International Institute for Environment and Development, agricultural machinery that is needed to raise the productivity of smallholder agriculture often requires more power than standalone renewable technology can provide. "The most effective interventions are those that take a technically neutral approach, giving preference to neither renewables nor non-renewables but having an objective view of what works best in a particular context".[lxi]

Dr Fatih Birol, the executive director of the IEA, has said that there is a real possibility that Africa might leapfrog the West by basing its economic development on renewable energy. There is certainly potential for this. Africa is rich in potential for PV, wind, hydro, and geothermal energy. However, these regions are also rich in coal and Africa is becoming a major source of oil and gas. Today, in many circumstances, fossil fuels provide the most accessible and lowest cost energy. Carbon emissions per person in Africa are only about one tenth of those in the rich, OECD, countries, so we cannot lay the burden of reducing

---

[lx] International Energy Agency, *Africa Energy Outlook*, 2014.
[lxi] Best, S. *Growing Power: Exploring Energy Needs in Smallholder Agriculture.* International Institute for Environment and Development, 2014.

emissions upon African shoulders. It is a race against time whether renewable energy, supported by smart grids, and perhaps also low-cost batteries, can become competitive soon enough to be the preferred options for the development of these regions. As always with economic development, stable government is the key factor. However, the willingness of the developed world to provide financial and technical help will also be critical.

12. Conclusions

Over the coming decades, to 2040, we will have to increase world energy by about a third, to meet growing demand in the developing world and to tackle energy deprivation. At the same time, we have to cut carbon dioxide emissions by 40% to be on track to limit global warming to 2°C. And this is only the start. Continuing steep reductions are needed to reduce emissions to near zero, or even to be net extractors of carbon dioxide from the atmosphere, by 2100.[lxii] We have to do this in a way that is affordable, bearing in mind the constraints on family, business, and government budgets in rich and poor countries. Climate change is a grave problem. We must find solutions that are economically and financially realistic.

In the immediate future improved energy efficiency will play an important part in moderating and reducing related emissions. This is by far the lowest cost option. We must work to overcome the inertia that resists this, in its many forms. But we also need a major contribution from technological change, which will be crucial to open the path towards a near zero carbon economy in the longer term.

It is already clear that wind and solar power, especially PV, are going to be major players. But we haven't yet overcome the problem of intermittency. We can see the challenges that European governments face as the share of variable renewables in electricity supply edges up, in 2016, to 15% in the UK and 20% in Germany. The jury is still out on how that will go. Spain, Portugal, and Denmark are doing better, but they have access to significant hydropower backup. Advanced battery technologies could

---

[lxii] IPCC WGIII, 2014. *Contribution to the Fifth Assessment Report, Summary for Policymakers.*

solve that problem but that would require big reductions in cost and clarification of the environmental implications. A 2016 study published by the UK Energy Research Council[lxiii] suggested that the costs of integrating variable renewables may be fairly modest at least up to a penetration of 30%.

In truth, we need to be reasonably cautious and recognise that we do not know the future. The "all singing, all dancing" option with renewables, smart grids, and flexible demand may provide the answer for some countries, we cannot yet be sure. Nuclear power and CCS also have their uncertainties. Hydropower and bioenergy will be important, but not all countries have them in abundance. The task is so challenging, and the uncertainties are so great, that we need to press ahead with development of all the main options.

We should continue to subsidise renewables but, as the technologies mature, the industry needs to become more competitive. The UK and Germany have already switched from open-ended fixed price guarantees to auctions for specific amounts of power. This is a good start. Support should reflect the value of the carbon emissions saved. As described in Chapter 9, we need a more systematic approach to incentivising carbon savings across the economy so that the full range of options are used. In most of Europe, subsidies that guarantee the price of renewables have insulated them completely from the electricity market, so that all the costs of intermittency fall on the other players and, ultimately, consumers. Renewables should be incentivised, relative to fossil fuels, according to the value of emissions saving. Beyond that renewables should earn the market price of the electricity they supply. In that way, the industry will be incentivised to engage with the problem of intermittency, for instance by teaming up with storage suppliers or with electricity customers with flexible demand. There are diminishing returns to continued investment in wind energy where there is already excess capacity at times when the wind is blowing. It is true, of course, that investors prefer a guaranteed price. Who wouldn't? But the industry is now reaching a degree of maturity where, with appropriate

---

[lxiii] Heptonstall, P. *et al. The Costs and Impacts of Intermittency — 2016 Update*, UK Energy Research Centre — 2016 update.

support for the carbon emissions that it saves, it can make its own way in the market place.

We should also invest in options for increasing the flexibility of power systems to accommodate high levels of variable renewables where this makes economic sense. These include expanding the grid, including international connections, and efforts to increase demand side flexibility. There is probably considerable scope for this at low-cost because, in the past, when we depended entirely on fossil power, time of day flexibility was not so much of an issue. It would be good to see more pilot exercises to test the willingness of consumers to sign service style power contracts and to respond to time of day incentives, because this remains one of the biggest areas of uncertainty. In due course, electric vehicles and electric heating of buildings may contribute to this flexibility, but in most countries mass adoption is still a long way off.

I don't believe that we will achieve low-carbon electricity without a contribution from reliable, dispatchable, low-carbon generation. Countries with hydropower already have that. France, which largely relies on nuclear power, has 90% low-carbon electricity and, as a result, emissions per person in 2015 were about half those of Germany and well below those of Denmark. Bioenergy may be significant, but probably the plant materials that are available in an environmentally acceptable way will always be limited.

Leaving aside hydro and bio, the two main candidates to provide reliable low-carbon power are nuclear power and fossil energy with emissions capture (CCS). Both have their problems. But if climate mitigation is a top priority, we must make every effort to overcome them. It is a questionable position, in my view, to say that one is committed to saving the climate while ruling out both nuclear power and CCS.

We have had one revolution, but now a second revolution is required. The first revolution has established renewable energy as a major contributor to electricity. The second revolution will require that we obtain virtually all electricity from low-carbon sources and that this low-carbon energy should become the major energy source for transport, homes, and industry. Much higher levels of energy efficiency will be needed to make this change affordable.

This second revolution will be politically more difficult. It includes the adoption of less universally popular technologies such as nuclear and CCS. Electricity consumers may have to enter into flexible "service" contracts to optimise the use of renewables. Motorists will have to switch to much lighter and less powerful vehicles and, probably, electric drive. Homeowners and businesses will have to insulate their buildings to a very high-level and, most probably, install heat pumps. Electricity, and energy in general, will be more expensive.

Not all of these measures need be unpopular. For instance, the electric vehicles will be quieter and less polluting, as well as cheaper to run. The time is right for city leaders to set specific timetables for the elimination of internal combustion engines for light vehicles within city limits. Cities can become much nicer places. I believe there will be public support. Cleverly designed service electricity contracts may also be quite attractive to consumers. Well designed, highly efficient homes can be very pleasant to live in, although existing homes present more of a problem. Energy may be more expensive but, with very high-levels of efficiency, individual bills may not go up by so much.

The transition may be costly but it is possible that once it has been achieved and the need to import fossil fuels has been eliminated or greatly reduced, energy costs may decline again. Hot countries, where the availability of solar power fits the pattern of demand for air conditioning, and where home heating is not such an issue, and countries with access to hydropower or large volumes of biomaterials may find the transition easier to achieve.

We need a broader and more honest debate about what the low-carbon transition really means. Governments obviously have a role in this, but this cannot be a "top down" exercise. The public must have a say and they must feel that they are positive participants. Environmental groups and informed citizens have a vital part in this. It may be easier for environmental organisations to stir up indignation against oil and gas companies and frackers, but these campaigns have no impact on actual carbon emissions. They could play a much more valuable role in influencing their own members, as well as the public at large, on the lifestyle changes that will make a difference. These include the steps that I have described; driving smaller, less powerful, and ultimately electric cars, insulating their homes, using public transport. They should also be working with the public and industry towards the flexible "service" contracts that will be needed if we are to make the maximum use of variable renewable energy.

Most of these conclusions are primarily directed to developed nations with the resources, capabilities, and responsibility to lead on climate change. In the developing world, as has been repeatedly emphasised, economic and social development is the first priority. In rural areas, where grid connection is not realistic, small island systems based largely on solar, wind and bioenergy make sense, sometimes with backup from diesel generation. The supply of clean cooking stoves is vital to reduce the scourge of indoor air pollution. However, economic development will require industrialisation and, as a general rule, urbanisation. By the late 2030s, most Africans may be urban dwellers. Huge investments in new energy will be needed to unlock the growth potential of these regions. There is a lot of potential for renewables. But on the IEAs central projections, a significant share of the new generating capacity in Africa and India will be coal or gas. Many developing countries have rich resources of one or both.

We need a close alliance between the governments of the developing and the developed countries, together with international financing institutions (IFIs) to promote rapid increases in energy supply along with clean energy transition. This is the heart of the matter, because all the growth in emissions is in the developing world and by 2040 some 70% of global emissions will be from outside the rich countries of the OECD. The delivery of off grid renewables in rural areas is certainly worthwhile but, on its own, does not get to the main problem. European nations and the US have campaigned to prevent the IFIs from investing in coal power stations. But this is coming at the problem from the wrong end. We do need to minimise the need for new coal plant, but this should be achieved in a less negative way by supporting investment in cleaner technologies.

Also vital is the promise that the developed world has been made to contribute $100 billion p.a. to developing countries in support of climate mitigation and adaptation. This is partly a matter of trust. If the West wants the developing world to pursue a lower carbon development route it will have to put its money where its mouth is. Finally, we need suitable international institutions for the developed and developing nations to engage on energy policy. I will return to this issue, and the need to reform international energy governance more generally, in Chapter 10.

We should take reductions in carbon emissions where we can find them. Ultra-efficient coal plant, switching from coal to gas, and gas CHP plant are part of the solution to the climate challenge in situations where the realistic alternatives would be worse. We should be working, progressively, to reduce the share of fossil fuels, without CCS, in energy supply. We deceive ourselves if we think that their use is coming to an abrupt end, so we had better make sure that new plant that is installed is as efficient and low-carbon as possible and we should upgrade existing plant where we can. CCS is our best option for squaring these investments with the eventually reducing emissions to very low levels.

And finally, research, research, research. The Paris climate summit gave new impetus to the effort to reduce climate change but it also demonstrated that the collective political will of nations is not sufficient to meet the 2°C target with existing technology. The key areas to work on include reducing the cost of renewables, especially PV, electricity storage, and reducing the cost and enhancing the effectiveness of CCS. We need to work on smaller, safer, and lower cost nuclear plant. Fundamental research in areas such as materials, chemistry, and biotechnology, may yield breakthroughs, but this is much more difficult to target. Most of the major nations at the Paris nuclear summit agreed to double their spending on clean energy R&D in the next 5 years through a new initiative, "Mission Innovation". That seems a promising start.

13. A Postscript: Negative Carbon and Geoengineering

What if we are unable to limit global warming to 2°C through the technologies described above for energy supply and use? The Paris summit has given welcome impetus to the cause of reducing climate change. But the fact remains that, as government policies stand, our efforts to reduce greenhouse gas emissions will not be adequate to meet the 2°C target, much less the new, tougher, 1.5°C target that was added in Paris.

So there is now increasing interest in alternative, in some cases radical, ways to cool the planet.[lxiv,lxv] These range from enhancements to existing efforts to reduce carbon emissions to geoengineering interventions in other aspects of the earth's temperature systems.[lxvi]

At the less radical end of the range we can move from reducing carbon emissions to actually extracting carbon that is already in the atmosphere. Indeed, in some of the more plausible "scenarios" for meeting the 2°C target we would temporarily overshoot permissible level of carbon dioxide in the atmosphere and then, when we have reduced the rate of emissions to a very low level, claw the position back by taking more carbon dioxide out of the atmosphere than we put in. This could be achieved by burning plant materials in power stations and then storing the emissions underground. The plants capture carbon dioxide from the atmosphere and we would store it after combustion. It is a fairly costly option which depends on the commercialisation of CCS, but it is much more cost-effective than the next stage, which would be to capture carbon dioxide directly from the atmosphere. This also is perfectly feasible, but it is much more expensive, and requires more energy, because the concentrations of carbon dioxide in the atmosphere are very low.

More radical options for reducing the carbon dioxide in the atmosphere include salting the oceans with iron or limestone to increase its CO_2 uptake. There are also proposals to increase the uptake of CO_2 that occurs naturally in the weathering of rocks. One way to achieve this would be to spread finely pulverised minerals on land. Since the product of weathering is alkaline and would eventually find its way to the sea, a secondary benefit might be to reduce the acidification of the oceans, a problem that other geoengineering interventions do not address.

Yet more radical are the options for reducing the temperature of the atmosphere of the earth by intervening more directly in the process of heating by the sun. We could spread aerosols in the atmosphere to reduce

[lxiv] Report of the Bipartisan Policy Centre's Task Force on *Climate Remedial Research.*

[lxv] The Royal Society, *Geoengineering the Climate*, September 2009.

[lxvi] National Research Council of the National Academies, *Reflecting Sunlight to Cool Earth*, National Academies Press, 2015.

the amount of sunlight reaching the earth, or we could enhance the reflectivity of clouds, possibly by spraying them with droplets of seawater. There are other geo-engineering possibilities, but these are the ones that have received the greatest prominence thus far.

We know that the aerosol option works because the material that major earthquakes throw into the atmosphere has been shown to have a cooling effect. There is also good experimental evidence, on a small scale, for the efficacy of the ocean seeding option. Compared to today's efforts to control global warming by reducing carbon emissions, these radical options would probably be inexpensive and those that directly reduce the solar heating of the atmosphere could be relatively fast acting, over just a few years.

These are measures of last resort because our knowledge of the atmosphere and of the oceans, which are enormously complex systems, is not nearly sufficient to understand the full consequences of these actions globally or regionally. Also, although they might counteract the increase of temperature, most of these interventions would do nothing to counter the other ecological consequences of increasing concentrations of carbon dioxide, notably the acidification of the oceans, which is a particular threat to corals and shell building organisms.

The issues are technical but also, some will feel, moral. Is it right to interfere with the planet in this way? There are also political questions. Who has the right to take such action and in what circumstances? What kind of international institutions or agreements would be needed to give such actions legitimacy? They certainly don't exist today.

Nobody is contemplating geoengineering interventions at present. They are not about to happen. The live question is, should we even be researching their feasibility and consequences? Some believe that by doing so we expose ourselves to "moral hazard" and that by raising the credibility of these deeply unacceptable options we only undermine the global resolve to reduce carbon emissions.

I agree that these are very unattractive and dangerous options. I doubt that researching them will reduce the commitment to emissions reduction because it is intuitively obvious to most people how risky they are. To some extent even talking about these options dramatises the magnitude of the challenge that we face.

But we cannot be sure of our success in emissions reduction and nor are we sure of the magnitude of the consequences of the climate change that is already inevitable. Because the future is so uncertain we need the best possible understanding of all available options, even those that it would only make sense to consider as the very last resort.

Chapter 7

Energy Finance, Fossil Fuel Subsidies, and Carbon Pricing

1. Introduction

A 20th century British politician, Rab Butler, was famous for saying that politics was "the art of the possible". But it would be just as true to say that finance is the art of the possible. The challenges of energy policy are financial just as much as they are technical and political. Governments, industry, and the financial community will need to be willing to adapt and to adopt forward looking policies in order to finance the transition to a secure, inclusive, and sustainable energy economy. If they get it wrong, the costs in terms of energy deprivation, wasted investment, and climate disaster, can be immense.

In this chapter, I outline the ways in which energy is financed today, the changes needed to meet energy policy objectives, and how the main players will need to adapt.

2. World Energy Investment Needs

Energy plays a large part in global investment. According to Deloitte, between 2009 and 2013, 27% of all capital raised was in the oil and gas sectors alone.

It may come as a surprise that, over the period 2010–2015, nearly 70% of energy supply investment was still in fossil fuel production and use, a ratio that had not changed much since 2009.[i] About $1.1 trillion p.a. was invested in fossil fuels, compared to $280 billion in renewables, $229 billion in electricity networks, and $13 billion mainly in nuclear and carbon capture and storage (CCS). Investment in energy efficiency was $220 billion p.a. So although renewables account for a large share of investment in new electricity generating capacity, they still account for less than 20% of energy investment as a whole.

In the IEA's projection for the period 2016–2040, assuming that governments deliver their stated climate policies, there is no dramatic change in the balance of supply-side investment and fossil fuels continue to dominate. The main change is a big increase in investment in energy efficiency. In contrast, to meet the 2°C climate target, the International Energy Agency (IEA) project that fossil fuel investment needs to approximately halve and investment in renewables to double, leaving the total level of investment in supply almost unchanged. However, also according to the IEA, investment in energy efficiency needs to increase nearly six-fold, underlining the vital importance of the energy efficiency of vehicles, buildings, plant and equipment, for climate mitigation.

A large part of projected investment in fossil fuels is in the US, the Middle East, Latin America, and Africa. Investment in new power plants, mainly renewables, is concentrated in China, Europe, the US, India, and other developing nations.

So achieving a transition to low-carbon energy while also meeting energy needs for economic development is going to require a major shift in energy finance. Fossil energy, especially oil and gas, will still be a big player in 2040, but there needs to be a big change in the balance of energy funding away from fossil energy and towards the electricity sector and especially renewables. There also needs to be a big increase in spending on energy efficiency, a topic that is discussed separately in Chapter 8.

It is not really a question of whether this money exists. Compared to world GDP of more than $72 trillion in 2014, these are not impossible sums. Finance is all about the management of risk. That is what

---

[i] International Energy Agency, *World Energy Outlook*, 2016.

determines the price that investors demand to commit their funds. The US government can borrow $18 trillion at an interest rate of 1.54% p.a. (for loans of 10 year duration). That is because investors have a high degree of confidence in the "faith and credit" of the US government. Most energy investments are a lot riskier than that.

3. Managing the Risk

In 2017, the world seems to be recovering from the financial crisis of 2008. Interest rates are low but financial institutions are more heavily regulated and cautious in their lending strategies. Many people have pointed to the need for governments to boost investment to lift the economy. The investment needed to meet energy policy goals could provide such a boost.

If we had a risk-free source of energy with the certainty of reliable continuing earnings then, financing it would be as easy as selling US government bonds. But in fact, energy investments are full of uncertainties. Will construction be completed on time and to cost? Will the plant perform as expected over its planned life? Energy prices are volatile. How will the prices of the product, and of any necessary feedstock, vary over this period? How might changes in government policy and regulation affect the economics of the project over its life? How secure are political relations with an overseas host government, where that is relevant? In the case of exploration and production companies, these uncertainties include the geology of underground formations that may or may not contain economic reserves. These are all risks that energy investors may face.

Energy finance is about packaging these risks into financial products that specific investors will find attractive. The ideal is to produce a sufficiently reliable flow of earnings that very large risk-averse lenders, such as pension funds and insurance companies, are willing to lend at low levels of interest. Pensioners and policy-holders must be able to rely on these earnings. There are many finance mechanisms including bank lending, project finance, public and private bonds, public and private equity, investment funds, and venture capital funds. Some of these sources may be able to accept a higher degree of risk, but they will require correspondingly higher rates of return. The sources of finance that are more willing to accept risk may represent a fairly small proportion of the total, but they play a disproportionate role in making projects possible and especially in financing adventurous, cutting edge initiatives.

However, in many aspects of energy finance, it is the willingness of governments and their regulating agencies to guarantee returns that is the dominant factor.

4. The Role of Government

Governments play a huge role in energy finance. In many developing countries some or all of the energy sector consists of government-backed companies whose earnings are subject to government price regulation. China and India are both examples. Outside investors are, therefore, largely betting on the reliability of government institutions and their willingness to maintain an attractive investment environment.

However, even in developed economies, governments have always had to regulate parts of the electricity and gas industries, that are judged not suitable for competition. This usually includes physical transmission and distribution of electricity and gas, and sometimes also generation. In the best cases this guarantees a fair rate of return, enabling the industry to be financed through bond and bank lending at low-cost. In the worst cases, political interference may lead to unsustainably low prices which undermine the performance of the industry. This has been a particular problem in India.

In the last few decades, European governments have gone beyond this to guarantee the price of renewable energy and also to set quotas. The justification has been partly that these were developing technologies that needed special support to break into the market and partly the overriding need to raise the share of low-carbon generation as part of national strategies to combat climate change. Guaranteeing prices in this way is a very powerful intervention and it has generally been very effective in boosting the market for renewables, but it also constitutes a big transfer of risk from the industry to energy consumers.

Even in the oil and gas exploration and production industries, traditionally viewed as private enterprise in the raw, government policies have a huge impact. Almost all major producing countries outside of Europe and the US have national oil companies (NOCs) with monopoly production rights which are to a greater or lesser degree the instruments of government. These companies are now by far the biggest producers. This arrangement is sometimes criticised as "resource nationalism", but it is not surprising, or unreasonable, for otherwise poor countries with a hugely valuable natural resource to want to keep its exploitation under close control. The problem arises when ruling elites misuse the resulting revenues, giving rise to the "curse of natural resources". The recent scandal in which more than $1 billion of oil payments seems to have been misplaced in Nigeria is only one example of many.[ii] But this is not necessarily the result.

Where private companies, such as the major oil companies EXXON, Chevron, BP, Shell, and Total, are involved the terms are set by host governments. It makes sense for governments that are taking a substantial share of profits also to share in the risk, since otherwise developments that could be worthwhile for both parties are likely to be inhibited. For instance, exploration and development costs may be allowed as deductions from the taxable earnings of already producing fields. Governments may adapt the severity of their tax regimes, from time-to-time, as oil or gas prices fluctuate. Where the government takes its rent in the form of production sharing, this may only cut in after a certain degree of profitability has been achieved for the developing company. Such policies have a big impact on the viability of new investment, as viewed by the financial community.

As a general rule, the effect of government intervention is to transfer business risks from project developers and investors to citizens, energy consumers, and taxpayers. Sometimes this makes perfect sense, for instance where there is no realistic possibility of a competitive market in a commodity that citizens require. It may also make sense where the government has a clear policy, in the national interest, that legitimately overrides commercial judgements. One example would be a policy that a certain percentage of renewable energy is required for climate reasons. There is a case for intervening directly to give certainty to the market and

---

[ii] Financial Times, *Unanswered Questions on Nigeria's Missing Oil Revenue Billions*, 13 May 2015.

ensure the lowest possible interest cost. However, in such cases it is vital that such intervention should be as market friendly as possible.

The role of the financial community, and of investing business, is to assess risks, promote economic efficiency, and allocate resources to where they will make the best return. Government intervention may cut across that role. For instance if the government guarantees the price of renewables or nuclear power, the financial community will play no role in estimating the value of the energy produced. The whole onus rests on government officials. They may have considerable ability. But they do not necessarily have the most relevant commercial experience. Investors do not need to have any knowledge of energy markets. The rate of interest required should be lower but there is no analytical contribution from the investing community. In some cases, where the government has strong reasons for wanting to promote new technologies, this sort of intervention may be necessary but it is a balancing act. In an extreme case, where the government takes on virtually all of the commercial risk there is an argument for financing new projects directly through government debt, which has the lowest interest cost of all.

Another drawback of government intervention is that inconsistency of government policy may become an investment risk. Put not your trust in princes. The problem is not necessarily that governments are perfidious but that policies may prove unsustainable, especially as they have not been tested against commercial criteria, or, indeed, that new governments with different policies may come to power. Some governments are considered more reliable than others. Spain has had to retrospectively remove financial incentives that had been promised to investors in renewables, when the uptake proved much greater than expected and the cost burden became unsustainable. The UK government has not rescinded any of its commitments to existing projects, but it has had to make sharp changes to schemes affecting new projects, which affects investment planning in the industry. Governments have now learned the lesson that such subsidy schemes should not be open ended and need to be capped to prevent runaway liabilities if they are too successful.

So far, we have mainly been considering support for domestic energy industries. International government support, intended to help poorer countries to reduce their energy related emissions or accelerate economic development, also has a major role in supporting necessary investment in developing countries with limited financial, technical, or administrative capabilities. These investments may be supported by international aid, or by export banks that guarantee the credit of host countries.

5. Risk Management Strategies

There are other ways of managing the risks of energy investment. Vertical integration is common in the energy industries. The International Oil Companies (IOCs) and big utilities are often retailers as well as producers. If the wholesale price of oil, gas, or electricity goes down, the margins at the retail end may improve, especially if there is a degree of monopoly, or pricing power, in the market. Consumers and politicians grumble, but to some degree that is a fact of life. Very large companies, such as the IOCs, have extensive investment portfolios and different relations with resource rich countries that enable to diversify their risk. They have technical and political expertise that should help them to assess and manage the risks effectively. Often very large oil and gas projects, though operated by one major developer, are financed and owned by a number of companies, thus sharing the financial risks between them.

Investors manage risk by diversifying their portfolios. Indeed, according to finance theory, risks that are particular to an industry or project and do not correlate with overall market movement, can be entirely diversified away. In a sufficiently varied portfolio, these risks are averaged out. On that basis only risks that are correlated with overall market movements merit higher than average returns. However, the rule seems to have limited application in practice.

6. Oil and Gas

90% of new investment in oil and 70% investment in gas is needed to replace the decline in production in existing fields. The industry has to run fast to stand still. That is why oil and gas investment will still be needed on a large scale even as world oil demand begins to decline. The biggest players in the oil and gas industries, and indeed in world energy investment, are the government-owned oil and gas companies. In 2016, 44% of world oil and gas upstream investment was by the National Oil Companies (NOCs), 24% by the "major" IOCs, and 32% by all others, including the US independent sector which was growing rapidly until oil prices fell in 2014.

By far the largest reserves of oil are held by the big Middle East NOCs such as Saudi Aramco and the National Iranian Oil Company. Their production is highly profitable even at relatively low prices. But their finances are closely tied up with national budgets. To varying degrees, their earnings represent the major source of government revenues. In principle, this could put pressure on their ability to invest at times when oil prices are low and government are under financial pressure. But their level of debt, or gearing, is generally rather low and they retain the ability to borrow because their production is still inherently profitable. They also have options for entering partnerships with IOCs to develop their reserves. Generally speaking, the NOCs have been able to maintain their investment even during the period of relatively low oil prices during 2015 and 2016.

In the longer term, decarbonisation to meet carbon targets represents a threat to the future of these companies and, and in the case of those most dependent on oil revenues, to the economic and political stability of their host countries. They may be able to continue producing as oil demand begins to fall, because their costs, at $10 per barrel or below, are so low. But as higher cost sources fall out of the equation prices will fall and the governments of low-cost producer countries will no longer be able to extract the huge rents available today. This might suit the importing countries, but the geopolitical consequences of undermining the economies of countries such as Saudi Arabia, Iran, and Iraq, are alarming.

Saudi Arabia is already trying to diversify away from oil through its National Transformation Plan. This includes selling off a 5% stake in its NOC. Depending mainly on the price of oil, the whole company may be valued in excess of $1 trillion, making it the most valuable quoted company in the world. The Plan also includes the establishment of a Public Investment Fund which, with other assets, could be worth a total of $3 trillion. The aim is to create non-oil investment income, to establish non-oil businesses and jobs and to modernise the economy more generally. It's an ambitious plan and it remains to be seen how successful it will be.

Other Middle Eastern producers, such as Iran and Iraq, with more limited resources, do not have the luxury of going down this route.

Historically, world oil production was dominated by the "major" private sector international oil and gas companies, such as EXXON, Shell, Chevron, BP, and Total. But the scene has changed radically in recent decades. By 2012 the "majors" accounted for only 12% of world oil production and 9% of world gas production. Other privately owned companies have bigger shares (24% and 19%), but by far the biggest producers are the NOCs of the Organisation of the Petroleum Exporting Countries (OPEC) and other producing companies (51% and 29%). The majors are, nevertheless, still amongst the world's biggest companies, with a total market value of around $1 trillion.[iii]

International oil and gas exploration and production is an extremely risky business. It requires very large investments in single projects and the deployment of advanced technology in unique situations. Even with the advanced seismic techniques that they now employ, the companies do not know for sure how productive geological formations will be until they explore and test them, and in their international operations they may be dealing with unstable regimes. All this, on top of the extreme volatility of oil prices, makes for a tough business environment.

It is the huge-scale of these companies, and the diversity of their operations, that enable them to convert large portfolios of such inherently risky projects into a relatively steady stream of income. One could say that that is their essential role. The shares in the major oil companies are held by pension funds, retirement accounts, and individual investors, as well as asset management companies. They largely fund their operations from their own massive cash flow.

The price of oil collapsed in 2014, from levels in excess of $100 per barrel to less than half that, and remained low through 2016. Yet the majors have weathered the storm, generally maintaining their dividends to shareholders. Share prices have declined, but not catastrophically. How has this been possible? One reason is that the majors are involved at both ends of the market. They retail oil as well as produce it. Lower oil prices are good news for their refining and marketing businesses. Also, the

---

[iii]Paul Stevens, *International Oil Companies, the Death of the Old Business Model.* Chatham House Research Paper, May 2016.

arrangements in which they share the profits from oil production with host countries are generally structured to limit their "downside". The companies are allowed to recover their costs before the "upside" is shared with the government. Nevertheless, by end-March 2015, all the majors had announced sharp cuts in capital spending. This reflected the fact that projects may not be so profitable with a lower price outlook, but it also helped to protect their cash position.

Some might say that it is an excellent thing to cut back on oil investment because that will ultimately reduce carbon emissions. That would be logical if there was a reasonable prospect that oil demand would also decline. However, on current government policies and in the light of consumer behaviour today that seems unlikely. It is more likely that we will witness another turn of the investment cycle in 5 or 10 years' time, with very high oil prices, significant damage to the world economy, and another investment boom. I hope I am proved wrong. To reiterate, most of the investment today is needed to replace the declining production from existing fields, not to meet increased demand.

How vulnerable are the majors to climate policy? As I have discussed in Chapter 6, we must hope that government policies and the habits of the motoring public will lead to a decline in oil consumption. With current polices, oil consumption will continue to rise at least for the next 25 years. The IEA judge that, to meet the 2°C target, it will need to have declined by 20% by 2040. Could the majors weather that? Their proven reserves cover only 10–15 years of production and their committed investment projects cover less than that.[iv] The majors have had to visit some very challenging locations to maintain their reserves and some of their more costly development projects are already being cancelled or delayed. Much will depend on what happens to production from the NOCs who dominate the market and what happens to the price of oil. Demand for gas continues to rise to 2040, in the IEA's 2°C case, so the companies may put more resource into gas. To some extent that is already happening.

If we succeed in limiting climate change to 2°C, the major oil companies, as we know them today, are eventually doomed. Successful

---

[iv] John Mitchell *et al. Oil and Gas Mismatches: Finance, Investment, and Climate Policy.* Chatham House Research Paper, July 2015.

deployment of carbon storage could, perhaps, improve that outlook somewhat. It makes sense, as Mark Carey, the governor of the Bank of England, has argued, that they should explain their exposure to low-carbon policies in their balance sheets. But if the companies are well managed it seems quite possible that their decline, over several decades, can be quite profitable and need not precipitate a financial crisis. The companies could, of course, seek to diversify into other more forward looking industries, such as renewables. But the business model is very different. Shell's diversification, in the past, into nuclear power, and EXXON's into electronic office systems, have been ill fated. BP's efforts to rebrand itself as "Beyond Petroleum" and to move into renewables in a big way also turned out to be unsustainable and the company has since drawn in its horns. Experience with the diversification efforts of major oil companies thus far has not been good. Probably the shareholders will prefer to have their money back. They will be very happy with that.

Today the major oil companies all maintain generous dividends and their stock market valuation is largely based on that, rather than on growth expectations. In February 2017, *The Financial Times* quoted Paul Sankey of Wolfe Research as saying, "the credibility of the management is reflected in the dividend yield ... shareholders just want their money back so they can put it into Tesla or whatever they think is next".[v] So you could say that the process of gradually transferring funds from big oil to renewable industries of the future has already started.

The independent private sector oil industry is, generally, less able to internalise its risk management and, accordingly, is much more dependent on outside sources of finance. High-risk exploration companies depend on equity investors or venture capital funds willing to trade risk for the possibility of big rewards. The biggest sources of finance are bank loans, bonds, and finance related to the earnings of specific oil projects. The very low level of interest rates in 2016 has meant that these instruments can be attractive even at low oil prices.[vi] The industry has suffered from low oil prices but there have been fewer casualties than expected from the sharp fall in oil prices in 2014 as the costs of oil and gas services and

---

[v] Financial Times, 11 February 2017.
[vi] BBC, *Shell Corruption Probe: New Evidence on Oil Payments*, 10 April 2017.

equipments have also fallen. Also the rapidly growing US shale oil and gas sector has been relatively resilient because its operations, which consist of a large number of low-yield wells with rapid pay-back, are more flexible than conventional oil projects.

7. Conventional Power Industry Investment

The oil and gas industry is ultimately financed on the basis of revenues from products that trade on international markets. Electric power can only trade within the relevant power grid and, even more important, 90% of investment in the power sector is within a framework of regulated prices. This means that the structure of these regulations and the predictability and reliability of the regulating authorities, are critical factors in power industry finance.

It is difficult to generalise about the financial structures of power companies around the world. As the IEA have said, "projects are carried out by a large number of heterogeneous investors that pursue different business strategies, have diverse management styles and expectations, operate in distinct legislative and tax systems, and have various asset portfolios". But, it is possible to identify some distinct problems that arise in financing the investment required to meet the world's power needs in a sustainable way.

Most investment in new generating capacity has traditionally been funded by large integrated utilities, whether private or government owned, from their own retained earnings or new borrowing. Project financing, in which the security is provided by the earnings stream of the investment rather than parent company guarantees, have also been common. The logic of these arrangements has been that big integrated utilities are best able to manage the risks because of their range of energy assets and their presence as electricity retailers as well as in generation. They have also enjoyed a degree of monopoly, or market power, although this is less openly acknowledged.

In some developed nations, particularly in Europe and to a lesser extent in the US, government programmes to promote renewables have undermined the role of the traditional utilities. It can be argued that the utilities themselves have been slow to recognise the importance of renewables. But in some cases, the policies for promoting renewables have been designed to attract new players. For instance in Germany, local individuals and communities have been encouraged to invest in wind and solar capacity, a policy that has helped to sustain local support. In much of Europe renewables have been guaranteed fixed "feed in tariffs", a policy that frees investors from any real engagement with energy markets.

Because renewables have very low variable costs, and may enjoy fixed prices whenever they run, they will inevitably run ahead of conventional plant where there is surplus capacity in a competitive market. In Germany and some other markets the regulations also give renewables priority. As the share of renewables increases it forces conventional plant to run sporadically as back-up when the renewables are least available. This causes a huge loss of value to the utilities who own the conventional plant. The combined income of the 20 largest publicly listed utilities in the EU fell by 85% between 2009 and 2013 and there was a big loss in their market value. The result has been a rapid retirement of older plant and almost no investment in new capacity. As a result, the German government had to agree in 2015 to pay utilities owning lignite power stations to keep them available as a reserve, at least until 2020. The UK government has instituted "capacity payments" to ensure sufficient capacity to meet winter demand peaks. In 2015, as described in the previous chapter, this mainly led to the commissioning of new diesel generators.

The weakening of traditional utilities is often represented as the triumph of new "disruptive" technologies and a positive sign of the coming low-carbon revolution. The assumption is that they will be replaced by much more nimble "service companies" offering sophisticated supply contracts that will contribute to the demand-side flexibility needed to cope with a larger reliance on variable renewables. This is by no means certain. As governments become less willing to guarantee the price of renewables we may eventually need the financial capacity of strong utilities to sustain high-levels of investment. The jury is still out on the question of how willing customers will be to adopt new contracts with flexibility incentives. In the process of change the existing utilities have the significant advantage that they know their customers.

The problem of how to finance back-up capacity will become increasingly challenging as variable renewables take increasing shares of electricity markets. This was discussed in more detail in Chapter 6 on energy technologies. Part of the solution, advocated in this book, should be a more realistic way of subsidising renewables, in which we pay them the value of their carbon savings, or tax other sources according to their emissions, but otherwise let their power trade in the market.

However, the main story in the developed world is the financing of the renewables themselves, which are expected to account for about 60% of all new capacity to 2035.

8. Renewables

A quarter of renewables investment in 2015 was in small distributed projects, such as rooftop and small ground-mounted solar PV.[vii] These may be financed by individuals or small communities. There is a proliferation of leasing financial mechanisms in which homeowners pay nothing upfront, entering into a long-term contract to purchase the power from the provider, usually at well below the standard utility rate.

The economics of small-scale PV continues to depend, in most countries on continuing government subsidies. These have been cut-back sharply in countries such as the UK, Germany, and Spain but remain in place, for instance, in Japan. The economics may also depend on regulations in a number of countries that require utilities to purchase excess electricity at their retail rates, known as "net metering". As distributed solar becomes more important, this rule is increasingly being questioned on the grounds that solar owners are not contributing their fair share of the costs of the grid. In the US, more than half the states that offer net metering have enacted changes or are considering doing so.[viii] However California, which has by far the largest small-solar capacity, has elected to retain retail pricing.

[vii] *Global Trends in Renewable Energy Investment 2016*, Report of the Frankfurt School-UNEP Collaboration Centre for Climate and Sustainable Development, in collaboration with Bloomberg New Energy Finance, for UNEP.

[viii] Report of the Frankfurt School, *Ibid.* Reporting the North Carolina Clean Energy Centre, which tracks this policy.

The decision by many European governments to guarantee the price of energy from renewable projects for 20 years or more represents a massive intervention. It transfers the market risk to the consumer. Even if the power from the project turns out to be worth very little, the consumer still has to pay the guaranteed price. Paradoxically, it may also have increased the regulatory risk, because investors are exposed to the possibility that governments may renege on their guarantee.

Governments took this step because they were convinced that it was necessary to achieve the rate of growth in renewables that they considered necessary to combat climate change and to meet national targets that they had set. In the UK, a more market-based approach was set aside when it failed to deliver the required investment.

The policy of price guarantee has been effective and has achieved the highest levels of renewables penetration in the world, in Denmark, Germany, and the UK. It has kept the interest rate required by investors down. And it has broadened the range of investing institutions beyond the traditional utilities. Most large-scale renewables projects are financed in ways that are not very different from other big energy projects. Large institutions employ a mixture of their own retained earnings, bank debt, and corporate bonds. The reliability of the project's earnings may make it easier to secure project finance in which the debt is secured by those earnings rather than the credit of the lead company.

Some special instruments have arisen. It has become a common practice to sell on major renewables projects, when construction has been successfully completed, to risk-averse new owners who value the reliable income stream. These organisations, known as "yieldcos" in North America and "quoted project funds" in the UK, raised more than $14 billion of equity between 2013 and 2015.

Overall, quoted US companies specialising in renewables have had a mixed experience in recent years. The WilderHill New Energy Global Innovation Index (NEX) has been volatile in recent years and has never fully recovered after it crashed from its peak in 2008. The most spectacular story was that of Chinese solar company Hanergy Thin Film, which rose meteorically to a value of $42 billion in late April 2015, before trading was suspended, as questions were raised about the quality of earnings. However, it resumed trading later in the year. In 2015 investment in solar energy in public markets was rising, up 21% on 2014 to just over $10 billion, while investment in wind was falling, down 69% to $2 billion, less than a fifth of its peak in 2007.[ix]

The scale of venture capital and private equity investment in renewables in 2015, although slightly up on 2014, was a shadow of its former self at 3.4 billion in 2015, compared to 9.9 billion at its peak in 2008. This kind of high-risk finance can facilitate new start-ups and provide a cushion to create lower risk investment opportunities, so its impact is greater than the absolute amount would suggest.

The UK government set up a Green Investment Bank in 2012 to stimulate investment mainly in renewables. Since then the bank has committed £2.7 billion to projects valued at a total of £10.9 billion. Institutions of this kind can have a catalytic effect by taking the lead in analysing and supporting green investment opportunities that others may join. The bank seems to have been quite successful in this role, although environmentalists have called for a much larger scale of operation. In 2017 the Government announced plans to sell the Green Investment to private investors.

9. International Government Finance

The Copenhagen climate summit of 2009 established a Green Climate Fund to finance climate mitigation and adaptation projects in developing countries. Membership of the governing board is split equally between developed and developing nations. As discussed in Chapter 4, The developed countries promised to "mobilise jointly $100 billion a year by 2020 to address the needs of developing countries. The funding will come from a variety of sources, public, private, bilateral and multilateral". A "significant portion" of the funding was to flow through the climate fund.[x]

Not surprisingly, this has led to continuing controversy as to what sources of finance governments are allowed to count against this total, and the precise meaning of a "significant" share. In June 2015, amidst preparations for the Paris climate summit later that year, Brazil, China,

---

[ix] *Global Trends etc, Ibid.*
[x] UNFCCC Copenhagen Accord, 18 December 2009.

India, and South Africa issued a joint statement of their disappointment at the lack of any "clear roadmap" towards raising the funds.[xi] They also criticised the lack of plans for "substantially scaling up" the support after 2020.

The developed countries issued their own definition of what counts towards the $100 billion in September 2015.[xii] Besides contributions to the Green Climate Fund this includes all climate funding by international financial institutions backed by developing nations, whether concessionary or not, plus export finance and development aid as well as private finance "mobilised" by this support.

Based on this definition, the OECD has estimated that a total of $61 billion was mobilised in 2014, up from $51 billion in 2013. The 2014 total was made of bilateral public finance (23.5 billion), multilateral public finance ($20 billion) and "mobilised" private finance ($17 billion).[xiii] The OECD are projecting that the total for public finance will rise to $67 billion in 2020, made up of $29.5 billion from multilateral finance and $37.3 billion from bilateral finance. They cannot predict the level of mobilised private sector finance. Based on these numbers one could say that the developed nations are making reasonable steps towards meeting their target in 2020. However, as of September 2016, the Green Bank had received only around $10 billion, including $3 billion from the US, $1.5 billion from Japan, $1.2 billion from the UK, and $1 billion each from France and Germany.

In the view of developing nations most of this money is simply rebranding of existing international development funds. The share of these contributed through the Green Climate Fund, where the developing nations share in control, remains rather small. Perhaps more significantly, the IEA estimate that developing nations, not including China, will have to invest at the rate of around $800 billion in renewables, electricity

---

[xi] The Guardian, *Rich Countries' Promise to Fight Climate Change "not delivered"*, 29 June 2015.

[xii] *Joint Statement on Tracking Progress Towards the $100 Billion Goal.* Paris, 6 September 2015 (18 donor countries plus the EU Commission).

[xiii] OECD, *Projections of Climate Finance Towards the USD 100 Billion Goal.* Technical Note 2016.

supply and distribution, and energy efficiency, between 2014 and 2035. Even $100 billion p.a. of international finance, which is for adaptation as well as mitigation, doesn't seem like a very big share of that. More recently, Donald Trump's announcement, in 2017, that the US will withdraw from the Paris Climate Agreement cast a shadow over the $100b promise, since the US was expected to be the largest single contributor.

One of the big issues here is the question of control. The developed nations have tended to channel their support through bilateral arrangements and established international financial institutions because that way they retain a large measure of control over the selection and the management of projects. To some extent that is reasonable because these are the tried and tested systems for managing investment projects in what are sometimes quite difficult environments. They need to be able to account to taxpayers where their money has gone. On the other hand, if the aim is to enlist the support of developing countries for the low-carbon transition, then these interventions will have to be made, to some degree, on their terms.

Plainly the rich nations are a long way from meeting their pledge, at least as far as contributions to the Green Climate Fund are concerned, as the 2020 target date draws near target date. Leaving aside the Fund itself, there are many routes through which developed countries contribute to climate-related projects in developing countries.

International financial institutions, such as the World Bank, the European Investment Bank, the Asia Development Bank, the European Bank for Reconstruction and Development, and the African Development Bank, are major funders of clean energy projects. The new Asia Infrastructure Development Bank, promoted by China, is another example. So are national development banks, such as KfW of Germany, Brazil's BNDES, the Japan Bank for International Cooperation, and the Export–Import Bank of China. Together they provided $84 billion of clean energy funding in 2014. However, only $15 billion of this was from lenders mainly owned by OECD countries to developing countries.[xiv] Now, increasingly, there are government sponsored "green" banks, of which the UNFCCC's Green Climate Fund is the best known.

[xiv] *Global Trends etc, Ibid.*

10. The Energy Charter Treaty

The flow of international investment will be enormously important in funding the growing needs of developing countries and supporting their low-carbon transition. Some of this will be in the form of aid from developed countries. But this will not be nearly sufficient on its own. Large flows of commercial and quasi-commercial investment will also be required. To make this possible, investors will need to be confident of fair treatment by the governments of host countries and their regulatory agencies. This includes freedom from expropriation of assets, and from discriminatory taxation or other regulatory burdens. Fair access to transmission or pipeline grids is crucial for inland investments in fossil fuels.

The Energy Charter Treaty was established in the 1990s to give just this sort of assurance to international investors. So far, its influence is mainly felt in Eastern Europe and Central Asia. It is highly desirable to broaden its geographical coverage, ideally in East Asia and Africa. The role and potential of the Treaty are reviewed in more detail in Chapter 10.

11. Conclusions on Energy Finance

Energy finance is not quite what it seems. One thinks of venture capitalists and startups promoting advanced and renewable energy technologies. They may have a disproportionate effect in promoting change, but they are, in reality, a small part of the total. Most energy investment is still in the production or use of fossil fuels, and the great majority is based on the decisions of governments or public sector regulators. The private sector organisations that are able to take the initiative in unregulated markets tend to be very large businesses such as the big IOCs and major gas and electricity utilities.

In the oil and gas industry the immediate risk is of under-investment leading to price spikes and high energy costs within the next few years. In the medium to longer term it is to be hoped that the industry will have to accommodate itself to the success of global climate policies. I am fairly sanguine about the ability of the major IOCs to do this, although it is a legitimate area of investor interest. Of much greater concern would be the geopolitical implications of the drying up of oil revenues in the Middle East.

There is a huge need for investment in the energy economies of emerging and developing economies to meet rising energy demand and achieve low-carbon transition. International aid — including climate aid — international financial institutions and green funds can make important contributions. But they will not be sufficient on their own. Meeting this need will depend, in large measure, on the energy polices of host governments and the extent to which investors can be confident of fair treatment in their markets. Governments need to provide a framework of reliable, predictable policy to make this possible. Mainly this is down to the host governments themselves. However, the Energy Charter Treaty has potential to help to build confidence in international energy investment. Its role is discussed in Chapter 10 on global energy governance. So could the transparency initiative, discussed in Chapter 2 on energy and global governance.

China is not as dependent on international finance as other developing and emerging economies. For China, the major programme of energy market reform that has been announced will be critical. This has already been mentioned in Chapter 5. China has been studying the reforms that took place in UK energy markets in the 1980s. They probably will not go as far as the UK has done and ownership of the major participants is expected to remain in the hands of national or regional government. But they are aiming for better separation between the electricity generators, the transmission companies, and system operators. Also they are aiming for an electricity market that dispatches power on the basis of least cost. This should improve the utilisation of wind and solar generators because they have very low running costs. It remain to be seen whether it will also open up opportunities for private sector investment.

The governments of developed countries committed to climate mitigation face a difficult dilemma. The $100 billion that has been promised, especially if the aid content is small, will not be sufficient to set the developing world on a low-carbon pathway to meet their rising energy needs. But there is also rising concern on the part of citizens in the developed world about the financial burden of low-carbon policies. It is essential for them to be able to show that their international contributions have been well spent. In many cases these international contributions, provided that they are indeed well spent, will be more cost-effective than domestic low-carbon subsidies in terms of global carbon savings per dollar.

The critical question, therefore, will be the ability of the governments of developed and developing nations to work together to optimise the effectiveness and scale of international funding. This requires the assurance of substantial and sustained funding on the part of donor nations. This funding must genuinely support the agreed energy policies of the host country. The host country must pursue policies that give assurance of fair treatment to international investment and that will lead to low-carbon transition at a rate that is appropriate to the country concerned. To some extent, the major international financial institutions are already working towards this. Chapter 10 discusses in greater depth the crucial question of how we can enhance North/South cooperation on energy policy and the institutions that are needed to achieve this.

The developed countries are facing a transition in the way that they support low-carbon technology. As average incomes have stagnated, a certain amount of consumer fatigue has set in over the cost of subsidising renewables. The cost of renewables has come down sharply, tending to reduce the need for these subsidies. But on the other hand, the need to incorporate larger shares of variable renewables in the generating mix poses a new challenge which also has cost implications. Dropping subsidies for renewables too suddenly could be disruptive and wasteful. But, as discussed in Chapter 9, it makes sense for governments to pursue a market-based strategy in which support is more closely based on the value of carbon emissions saved. This is likely to lead to a more "normal" outcome, in which renewables are largely financed by major utilities as part of the development of their portfolios of generating assets.

12. Subsidies for Fossil Energy

Environmental groups and international bodies regularly claim that the fossil fuels — coal, oil, and gas — receive far greater subsidies than renewables.[xv,xvi] What is the true picture?

The huge coal, oil, and gas industries have complex relations with governments across the world involving many different payment streams.

[xv] Greenpeace Daily Dispatch, *G20 Fossil Fuels Get Way Bigger Subsidies than Renewables*, 15 November 2015.
[xvi] International Monetary Fund Survey, *Counting the Cost of Energy Subsidies*, 17 July 2015.

In most cases, oil and gas production is highly profitable and pays large rents, in one form or another, to government. In the UK, North Sea oil and gas production, during its most profitable phases, paid high rates of Petroleum Revenue Tax. The UK also has very high taxes on petrol consumption. At least half of the price that UK motorists pay is tax. Most US states charge special taxes, known as "severance taxes" on oil and gas production. In the most productive countries in the Middle East, revenues from oil and gas production contribute substantial shares of all government revenues. However, to varying degrees, governments offer deductions against these taxes for legitimate costs and to promote investment. In the UK the oil companies expect to claw back some of their tax payments to pay for the decommissioning of offshore rigs. Taxes may be reduced, and concessions made to promote new developments (such as shale gas in the UK) or to shield the industry when oil prices are low. I would not describe these concessions as subsidies, provided that they are in the context of overall tax payments that are higher than would prevail in the non-oil and gas tax regime. To oil and gas producers, therefore, it may seem odd to claim that their industry is heavily subsidised.

The coal industry is in a different category, because, typically, it does not generate such large economic rents. The coal mining industry in Europe, especially in the UK and Germany, has been heavily subsidised in recent decades, almost entirely for social reasons. Imported coal was cheaper, but the government was anxious to preserve jobs in the mining regions. Now the UK has virtually closed its coal mining industry and these payments have come to an end. But the government is still making legacy payments, on a large scale, to miners suffering lung disease. Similarly, the UK government is making legacy payments, of about £2 billion p.a., to manage the waste and decommissioning of nuclear power.

In the UK and Germany, which have amongst the highest levels of renewables penetration, government is providing incentives, intended to be temporary, to keep coal and gas power stations available to help cope with the intermittency of renewable energy supply. Is that a subsidy to the coal and gas industries? One could equally well argue that it is an indirect subsidy for renewables.

The IEA have estimated world fossil fuel subsidies in 2015 at $325 billion, compared to $120 billion of subsidies for renewables.[xvii] Most of the fossil fuel subsidies are the result of cheap energy policies in the big oil and gas producing countries, especially in the Middle East and some major developing countries. The IEA counts the difference between a domestic price and the world market price as a subsidy. The biggest offenders are oil producers in the Middle East, such as Iran and Saudi Arabia, plus India, Russia, Venezuela, Egypt, and Indonesia. To a lesser extent, major developing countries also subsidise fossil fuels.

It is definitely desirable to reduce, and if possible eliminate, the policy of selling oil or gas at below international market value. This practice is leading to rapid growth of oil and gas demand in the Middle East, reducing export revenues, and undermining government finances. It also undermines the market for renewables and the incentive for energy efficiency. It will be much more difficult, for instance, for Saudi Arabia to promote solar power, in which it has a natural advantage, as long as a cheap fossil energy policy persists.

But these habits are hard to change. Most people don't think like economists. Paying the production cost for a commodity that is plentiful and cheap to produce locally doesn't feel like a subsidy. The benefits fall mainly to the middle classes, although there are also benefits to the poorest. The middle classes may not be the most in need but they are often politically influential.

The G20 is committed to phasing out fossil fuel subsidies. Their 2016 summit *communique* said, "we reaffirm our commitment to rationalise and phase out inefficient fossil fuel subsidies that encourage wasteful consumption over the medium term, recognising the need to support the poor". There is plenty of wiggle room here. Which subsidies are "inefficient" and "wasteful"? What does the "medium term" mean? The text bears the hallmarks of a negotiation with countries who are not necessarily convinced that their energy subsidies are inefficient or wasteful and who are unwilling to commit to a timetable for phasing them out.

The G20 has also instituted a voluntary process of "peer review" for reducing fossil fuel subsidies. China and the US reviewed each other in

---

[xvii] International Energy Agency, *World Energy Outlook*, 2016.

2016, reflecting the positive leadership role of these countries in climate diplomacy. In their report the US confessed to $1 billion p.a. worth of subsidies, mainly in the tax code, while China confessed to subsidies worth $15 billion p.a.[xviii] However, the report does not contain specific plans to bring these subsidies to an end. Germany and Mexico are following.

Subsidies as measured by the IEA fell from close to $500 billion in 2014 to $325 billion in 2015. This was mainly the arithmetical result of the decline in oil prices. There is, nevertheless, steady progress around the world in reducing fossil fuel subsidies, and the decline in oil prices in 2014 has made this less painful.

There is not much overlap between the countries providing large subsidies for renewables and the countries with large subsidies for oil and gas. Generally speaking in the West, on a strictly financial basis, it is the renewables (and in the UK nuclear power) that are subsidised, whereas fossil fuels pay substantial special taxes. It is misleading to suggest otherwise.

Subsidies for fossil fuels are certainly a big barrier in the way of climate mitigation and economic progress in many parts of the world. At the very least we should be aiming for a level playing field between fossil energy supply, improved energy efficiency, and low-carbon alternatives. Reducing these subsidies is a no regrets policy that is good for the economy as well as the climate. But of course it is politically difficult to achieve. As the IEA have said, "subsidies are closely linked to sensitive political issues such as the sovereignty of governments to use natural resources as they see fit, trade competition and poverty alleviation".[xix] Nevertheless, the G20 initiative is important and, hopefully, will continue its steady progress.

There is another, quite different, way of looking at the question of subsidies. If one includes the right of the fossil fuel industries to freely vent harmful emissions into the atmosphere as a subsidy, a different picture emerges. On this basis, an IMF working paper published in 2015

[xviii] OECD Press Notice. *OECD welcomes ground breaking peer reviews by China and India of their fossil fuel subsidies*, 6 September 2016.
[xix] International Energy Agency, *World Energy Outlook*, 2014.

estimated total fossil fuel subsidies to be in the region of $5 trillion, about 10 times greater than the IEA's estimate which only included financial aspects.[xx] Of course, these environmental costs are hard to measure accurately. Nevertheless it is clear that, if they are included in the equation, the fossil industries are indeed far more heavily subsidised than renewables. This, of course, is one, rather compelling way of putting the case for action on climate change and local pollution. It points to the need to put a price on carbon emissions.

13. Carbon Pricing and Carbon Trade

Government policies to promote low-carbon transition can take a variety of forms. In his 2006 report, Sir Nicholas Stern divided these into three main categories. First, there are those that put a price on carbon emissions, so that energy markets take account of their inherent costs. These are the main subject of this section. Stern has described the absence of such a price signal as, the "greatest market failure the world has seen". The mechanisms are discussed below. Secondly, there are technology-specific policies. The economy cannot respond fully to carbon pricing signals unless low-carbon technologies are available. These policies include support for R&D, demonstration and deployment of specific low-carbon technologies. The guaranteed prices or "feed in tariffs" that many European governments have adopted for wind and solar are examples. The third category covers policies that are aimed at removing barriers to change. Without these the inertia of the system and the lack of awareness of opportunities may mean that worthwhile opportunities are not taken up, especially for improving energy efficiency. This category includes a wide range of initiatives affecting regulation, information, and finance, education, persuasion, and debate. Good examples of this include providing information on the energy efficiency of buildings and appliances, or installing "smart meters" to raise awareness of energy saving options.

So a range of policies will always be needed to reduce carbon emissions. However, carbon pricing and carbon trade should be at the heart of global energy policy. There are three overlapping reasons for this. Firstly,

---

[xx] International Monetary Fund Survey, *Counting the Cost of Energy Subsidies*, 17 July 2015.

they are the best hope of rebuilding a genuinely global approach to combatting climate change following the "bottom up" outcome of the 2014 Paris summit, in which progress was left to the individual offers of national governments. Article 6 of the Paris agreement covers the possibility of international emissions trading, but specific mechanisms are yet to be worked out. Secondly, they are essential for keeping the costs of climate mitigation under control and consistent with affordable energy. And thirdly, it is only through carbon trading that climate mitigation resources can be concentrated where they are really needed to finance low-carbon energy solutions for rapidly growing developing nations, especially in Asia and Africa.

According to the World Bank,[xxi] an international carbon market may reduce the cost of achieving national emissions targets by a third by 2030 and by more than half by 2050. "In the period beyond 2050 it is highly improbable that the emission reductions needed to meet a 2°C or lower target can be achieved without this flexibility. The earlier that an international carbon market is developed, the larger the savings and hence the greater the potential to scaleup ambitions in the short-term. The modelling analysis also shows that some of the poorer regions in the world may be able to generate financial flows amounting to 2–5% of GDP in 2050 through the use of an international carbon market."

Carbon pricing is an especially powerful policy for carbon reduction because, potentially, it brings all possible options into play and incentivises the full range of participants in the energy economy. Without carbon pricing, governments are at risk of pursuing specific low-carbon initiatives that are much more expensive than the true marginal cost of carbon saving in the economy. The principle applies even more strongly on the international scene. Without carbon trading the risk is that rich countries may pursue ever more expensive policies for reducing carbon emissions to near zero, while developing economies pursue industrialisation based on fossil fuels. Opportunities are missed for low-cost and larger scale carbon reduction, together with low-carbon relief of energy deprivation.

The mechanics of carbon pricing are fairly simple in principle, though they may become more complex in detailed application. A carbon tax is

---

[xxi] World Bank Group, *State and Trends of Carbon Pricing*, October 2016.

simply a government charge on an organisation that emits greenhouse gases that is calculated as a fixed amount per unit of emissions. Usually it just relates to carbon dioxide, but it can be applied to the carbon dioxide equivalent of all greenhouse gases. In some cases these taxes are just applied to the largest single point emitters, that is to say power stations and major industrial plants. Applying the tax to all emitters, including domestic energy consumers is more complicated, and perhaps more difficult politically, but perfectly feasible. Generally speaking it is politically more attractive to apply the tax to the energy producers, such as big power generators, than to consumers, such as you and me in our homes. Obviously, carbon taxes need to be comprehensive to be fully effective.

The main alternative, known as "cap and trade" is a bit more complex. The government has first of all to decide on the rate of emissions reduction that it is aiming for and, thus, the maximum emissions to be allowed in each period of time. The government sells, or in some cases grants, quotas equivalent to these amounts and makes it obligatory, in the sectors of the economy covered by the scheme, to surrender quotas equivalent to all greenhouse gas (or carbon dioxide) emissions. Because the quotas can be traded, this should lead to the least cost options for carbon reduction being adopted.

As discussed below, governments can also set internal carbon prices that are simply guidelines, to be used for their own policymaking, on how the harmfulness of greenhouse gas emissions is to be taken into account. This is likely to be much less effective than carbon taxes or cap and trade systems, but it helps to make environmental policies more rational.

So far, we have been talking about schemes put in place by individual national governments. There are also schemes that have been adopted by cities or regional governments. There have also been international cap and trade schemes, such as the European Emissions Trading Scheme (EU ETS) and the Kyoto Protocol scheme, in which national quotas are set by international agreement and quotas are traded across national boundaries. These schemes are discussed further below.

13.1. *Current Global Status*

According to the World Bank about 40 national jurisdictions and over 20 cities, states, and regions are putting a price on carbon through carbon taxes or emissions trading systems, and the numbers continue to grow. The best known of these, and the one with the largest coverage, is the EUs ETS. But there are many other examples. China has a number of large cap and trade pilot schemes already operating and has announced its intention to inaugurate a national ETS in 2017. Depending on its exact scope, that will increase the share of world emissions covered by carbon pricing from 13% today to between 20 and 25%, a substantial step forward. So this is not a neglected area.

Most of these initiatives are, however, rather half hearted. About three quarters are set at below $10 per tonne of CO_2, whereas the price of carbon needed to achieve the 2°C climate target has been variously estimated in the range of $80–$120 in 2030. And there is very little international trading. The Kyoto Protocol of the climate treaty (UNFCCC) provided for trading in climate credits and this took place on a large scale in its first round, up to 2012. But only the EU, amongst major players, endorsed a continuation to 2020, and trading under this "top down" mechanism has now largely dried up.

Some governments also set carbon prices as an internal guideline for policies relevant to carbon emissions, as described above. 17 out of 23 governments surveyed by the OECD had such prices. In 2014, they averaged $38 per tonne of CO_2 for energy investment projects.[xxii] But prices set for future years may be much higher. The average price for 2050 was $153.

Many companies have also adopted internal carbon pricing in their decision-making to promote low-carbon transition and to protect themselves from the impact of future emissions regulation. Over 1,200 companies reported to the Carbon Disclosure Project in 2016 that they were currently using an internal carbon price or planned to do so in the next two years.[xxiii]

Industry is a strong advocate of carbon pricing because it helps businesses to plan for their transition to a low-carbon world and avoid investments that may become stranded. In September 2014, a group of over 400

[xxii] Smith, S. *et al. Monetary Carbon Values in Policy Appraisal: An Overview of Current Practices and Key Issues*, 23 September 2015.
[xxiii] CDP, *Embedding a Carbon Price into Business Strategy*, September 2016.

investing institutions representing more than 24 trillion in assets called on governments to, "provide stable, reliable and economically meaningful carbon pricing that helps redirect investment commensurate with the scale of the climate change challenge".[xxiv]

13.2. *National Systems*

Carbon pricing subscribes, unfortunately, to the well-known dictum that the most effective policies for addressing climate change also tend to be, politically, the most difficult. Profound changes will always encounter tough resistance. But profound change is needed. Most energy politicians are well aware of the sad case of Australia's carbon tax, introduced by Prime Minister Julia Gillard in 2011. It was the subject of fierce criticism. Opposition leader Tony Abbott vowed to lead a "people's revolt" and to "fight this tax every second of every minute of every day".[xxv] His campaign was successful. Ms Gillard lost her job in 2013 and the tax was repealed in 2014.

Most carbon pricing initiatives have not had such a rough ride as this. But there is no pretending that carbon pricing is a politically easy option. To put this another way, it risks cutting across other objectives, such as affordable energy and jobs in carbon-intensive industries.

Governments should start by setting an internal carbon price as a guide to rational policy making and also as a signal to business. As with pricing instruments themselves, this should start low and rise progressively over time. Sudden introduction of a substantial carbon price has the effect of handing windfalls to existing low-carbon energy producers, many of whom will already have received government subsidies. The important thing is to incentivise new investment, and that is about prices in the future. So it makes sense to set a gradually rising scale.

The drawback is that business may not be convinced that future administrations will stick to the announced policy, especially if the going gets tough. That is a problem, but it can be overstated. Business people cannot predict such fundamental factors as the future price of oil or currency exchange rates. They are used to dealing with uncertainty. A clearly

---

[xxiv] *Global Investor Statement on Climate Change*, September 2014.
[xxv] New York Times, *A Carbon Tax's Ignoble End*, 24 July 2014.

stated government policy, especially if there is widespread political support, can send a powerful signal.

There is a large literature on the respective merits of carbon taxes and cap and trade systems. The advantage of taxation is that it is simpler and the cost to emitters is known. But the outcome, in terms of emission reductions, is uncertain. The advantages of cap and trade are that the level of emissions is known and that it opens up the possibility of international trading in permits. But the resulting price of carbon, and therefore the burden on emitters, is uncertain. Over an extended period the outcomes may not be so different, because governments will always make adjustments to the schemes that are in place if the outcome is different from what they intended.

The most significant carbon pricing policy in place today is the European cap and trade system, i.e. ETS. The price of permits has been much lower than expected, falling below 4 Euros (about $4) per tonne of CO_2 in 2016. This can be attributed partly to the economic recession following the financial crash and partly to politics of the process, in which each government has tried to protect its industries by ensuring the adequacy of permits. More positively, it has been due to the success of other low-carbon policies, notably support for renewables, in reducing carbon emissions. Some believe that the very existence of the scheme and the associated carbon accounting has had a powerful effect.

Whatever the reason, a carbon price of $4 is too low to have much impact on carbon emissions and sends an erratic signal to industry. The European Commission aims to adjust the system in future rounds to achieve a more meaningful price.

Interestingly, the UK government has adopted a "belt and braces" approach. The UK is part of the EU trading scheme, but in order to give a more consistent signal, in 2013 the government also adopted a "minimum carbon price", set at £25 (about $33) per tonne of emissions in 2017 and intended to rise progressively in the future. This takes the form of a tax on carbon emissions set at whatever level is necessary to make up the difference between £25 and the price of ETS permits. This seems like a practical, pragmatic approach to getting the benefits of both systems. However in 2014, facing concerns about the impact of the policy on utility bills and on the competitiveness of energy-intensive industries, the government froze the rate of the tax until 2021. This, unfortunately, rather undermines the intention to send a clear signal of steadily rising carbon prices to industry. It underlines the political difficulties of carbon pricing. However, more recently, France has announced the intention to set its own minimum carbon price, so perhaps this approach has a future.

We have already noted the political difficulties that surround carbon pricing. It raises utility bills, which are especially burden some on the poor, and it handicaps energy-intensive industry in international competition. It can be counter-productive if it causes these industries to migrate to countries with relatively lax environmental regulation.

However a positive factor is that, in contrast to policies for subsidising low-carbon options, carbon pricing raises revenue. These revenues can be used to ameliorate the negative impacts. Social support can be provided for those who may find it difficult to afford heating and lighting, and heavy industry may receive tax concessions or investment credits. In cap and trade schemes, free carbon quotas can be given to industries that genuinely face stiff international competition. Of course these measures cannot precisely remove the impact of the tax on individual businesses, because the whole purpose is to incentivise carbon savings, but they could make them impact neutral, or near neutral, across industrial sectors. There is an argument for devoting the proceeds of carbon pricing to carbon reduction measures, but the first priority must be given to the adjustments that are needed to ensure the continuing political viability of the policy.

Some have argued that, to protect domestic industries, countries with carbon pricing in place need also to impose a "carbon border tax" on the carbon content of imports from countries that do not have similar environmental policies.[xxvi] This would be a big mistake. The founding principle of the global climate agreement (UNFCCC) is that the rich countries must give a lead. They originated the problem of global warming in the first place and now have the resources needed to tackle it. The developing countries must also play their part but their first concern is to raise living standards and it was recognised that they would need more time. Imposing carbon taxes on the exports of developing nations would be a violation of

---

[xxvi] Financial Times, *Green Barricade; Trade Faces a New Test as Carbon Taxers go Global*, 23 January 2008.

this principle. It would disrupt their trade with the developed countries which, in many cases, is their best hope of economic progress. It is hard to think of a policy that would be more likely to undermine the hopes of cooperation between North and South on global warming. As I have described, there are better ways of protecting the interests of energy-intensive industries.

13.3. *International Carbon Trading*

There are two main possibilities for international carbon trading. The first is when two countries with national carbon quotas can exchange credits. So a country that is coming in below its quota can sell the surplus to a country that is overshooting. In fully developed cap and trade systems the trading can take place between businesses and not just at government level. The second is where investors in an overseas carbon reducing project in a country that may or may not have an internationally recognised carbon quota are able to claim the credit in their home market.

Both of these options were recognised in the Kyoto Protocol. The aim of the Kyoto Protocol was that developed countries would accept binding emissions targets but that developing nations would not, although this was never fully realised after the US declined to commit to its target. The Kyoto Clean Development Mechanism (CDM) enabled countries with quotas to claim credits arising from carbon reducing projects in countries without quotas, and the Joint Implementation Mechanism (JIM) enabled them to be claimed where both countries had quotas. Trading under the Kyoto mechanisms was substantial but, as described above, has now largely dried up as participation in the Protocol has declined.

International trading of carbon quotas, in any form, requires high level of mutual trust and appropriate regulation. Because the issue is so important, there is now a plethora of institutions seeking to promote carbon trading and the building of a suitable framework. These include the Carbon Pricing Leadership Coalition and the New Zealand led Ministerial Declaration on Carbon Pricing. There is also a High Level Panel on Carbon Pricing and a G7 Carbon Market Platform.

For different cap and trade mechanisms to correspond there has to be shared confidence in the regulation and transparency of each system and a recognition that the caps are being set at similarly demanding levels.

Otherwise there is a risk of what has become known as "hot air". This is where participants set unduly lax targets and are rewarded by being able to sell quotas that do not represent any real reduction in carbon emissions below business as usual. Full trading between cap and trade systems should lead to a single price for carbon credits across the systems that are included.

In 2013, the cap and trade systems of California and the Canadian province of Quebec were formally linked in this way, following a finding by the Quebec authorities that the California system was as "stringent and comprehensive" as theirs. No doubt it is a lot easier to achieve this level of mutual confidence between these two North American regions than it would be between countries at different levels of economic development. Nevertheless, it is an interesting example.

The new Chinese cap and trade system, which is supposed to come into operation in 2017, is based, at least in part, on the European model. Trading between these two systems would represent a major step forward in international carbon markets and provide a possible model for a global system. But no doubt European authorities will need to see how the Chinese system performs, and Chinese authorities will want so see a bit more price stability in the EU system, before this could happen.

Mechanisms that enable international low-carbon investments to generate carbon credits have to be able to demonstrate that the carbon savings are genuinely additional to what would have happened without the finance that the credits provide, always a hard thing to do.

International carbon trading mechanisms do not have to cover all sectors of the energy market. For instance the EU's ETS only covers power generation plus energy-intensive industry. There is considerable interest in the possibility of sector-specific carbon trading systems between developed and developing nations because, at least as a first step, these might be easier to put in place, and to regulate effectively, than more comprehensive arrangements.

14. Conclusion

Carbon pricing and international carbon trading are fundamental to combatting climate change. They are necessary to bring the full range of low-carbon options into play, to link the efforts of rich and poor countries, and to financing low-carbon growth and transition in the developing world. Unfortunately these policies are also politically challenging.

It makes sense for all governments to adopt internal carbon pricing on a rising scale to promote consistency and cost effectiveness in their climate policies. Governments, starting with the developed world, should also adopt carbon pricing instruments, whether in the form of a carbon tax or a cap and trade system, as many are already doing. There is merit in combining a cap and trade system with a steadily rising minimum price, as the UK has aimed to do.

Carbon pricing is not an elixir on its own and needs to be backed by other policies to promote new technology, and remove barriers to market response.

Carbon pricing must be implemented in ways that are sensitive to other energy policy objectives, notably industrial competitiveness, and the amelioration of fuel poverty. Otherwise its political viability will be in question. This must have the first call on the funds raised.

Governments should avoid imposing carbon border taxes because these are likely to be counter-productive for the global climate effort and for economic development.

After the decline of trading under Kyoto mechanisms we need to rebuild international carbon trading in a more "bottom up" way. There are numerous growth points represented by individual trading schemes and international institutions. Developing the framework provided by the Paris climate agreement is an important part of this. The development of trading relations between the European and Chinese cap and trade systems, would be a valuable step with global significance, if that turns out to be possible. Probably this will take a long time, and success may only be partial, but the ultimate outcome of global cooperation on carbon abatement is at stake.

Chapter 8

Energy Efficiency

1. Introduction

Most of the energy that we generate is wasted. The scale of this waste is breathtaking. Most of the energy generated at power stations is dissipated into the atmosphere, as is much of the energy used to heat our, mostly poorly insulated, homes. Much of the heat that is retained in our homes is wasted on empty rooms and, to demonstrate our status in the world, we drive cars that are much bigger and more powerful than is needed to get us from A to B.

At headline-level, improving energy efficiency is the most attractive way to solve the world's energy problems. Every generation of energy politician rediscovers this. Energy efficiency reduces pollution, mitigates climate change, reduces the pressure on natural resources and, at the same time, *saves money.* For energy importing nations it improves the trade balance. What could be better than that? The potential is enormous. In academic studies also, greater energy efficiency usually figures as by far the largest single contributor to the emissions reductions needed to protect the climate.

There is nothing new in this. For at least the past 40 years, energy efficiency has been hailed as a benign, and hugely under-exploited additional fuel. Governments have resolved to transform the energy efficiency of their economies. Faced with a range of "scenarios" of the future of energy supply and demand, politicians will always pick those with high rates of improvement in energy efficiency. But of course that does not necessarily mean that they will be able to adopt policies for bringing this about.

The steady historic improvements in energy efficiency have certainly made a big difference to todays' energy economy. For developed countries[i] energy efficiency improvements averaged about 2% p.a. over the period 1973–1990 and then dipped to about 0.9% p.a. between 1990 and 2005. This may seem a modest rate but without these improvements energy use in most of these countries would have been 58% higher in 2005 than it actually was. These long–term trends appear to underline the importance of energy prices as an incentive. In their study of 2007, the International Energy Agency (IEA) drew the conclusion that, "the changes caused by the oil price shocks in the 1970s and the resulting energy policies did considerably more to control the growth in energy demand and reduce CO_2 emissions than the energy efficiency and climate policies implemented in the 1990s".[ii]

However in 2014, and again in 2015, the rate of improvement in energy intensity improved again, to 1.8% p.a.[iii] The IEA concluded that this reduction in energy intensity made the largest contribution to halting the growth in carbon dioxide emissions in 2015. They attribute it partly to the increase in mandatory efficiency regulations but also to changes in industrial structure, especially in China. Energy intensity declined by more than 6% in China in 2015 and despite continued economic growth, total energy demand declined slightly.

This improvement in global energy intensity is undoubtedly one of the most encouraging recent developments in the effort to moderate climate change. However, the IEA estimate that, together with measures to decarbonise energy supply, global energy intensity needs to decline even more rapidly, at an average rate of 2.5% p.a. between now and 2040, to get us on track to meet the 2°C climate target.

---

[i] I.e., the members of the OECD.

[ii] International Energy Agency, *Energy Use in the New Millenium: Trends in IEA Countries*, 2007.

[iii] International Energy Agency, *World Energy Outlook*, 2016.

In spite of the recent improvement, the rate of improvement in energy efficiency still falls far short of what is possible. This is a paradox. How can so much rhetoric, and indeed considerable effort, to access a benefit that seems so accessible and so attractive, have failed to deliver the full potential? In order to understand this we need examine the role of energy efficiency and the associated policy options in a bit more detail.

2. Industrial Structure

Energy efficiency, properly defined, means reducing the amount of energy needed to achieve the same result. That is the subject of most of this chapter. But energy efficiency is not the only, or even the main thing that determines the overall energy intensity of an economy, in other words the ratio of energy to GDP. This varies enormously between different countries according to their stage of development and the balance of their activities.[iv] For instance developing countries, such as India or China, need the energy equivalent of a tonne of oil for each $2,000 of GDP, whereas developed nations produce more like $10,000 for each tonne of oil. The main reason for this is not that the developed nations are so much more efficient, although that is certainly part of the explanation, but that they have more mature economies that are less dependent on heavy industry. In contrast, developing countries are rapidly building their infrastructure of roads, houses, schools, sewage, and power systems, and the like, all of which require masses of energy.

Take China as an example. China's spectacular economic growth of recent decades has been built on the export of manufactured goods and, as a result, manufacturing accounts for almost one third (32%) of its economy, an extraordinarily large share by international standards. For most other countries the share is in the range of 10–20%, for example, 10% for the UK, 13% for the US. Domestically, China, as a developing nation, is building its infrastructure of roads and buildings at an extraordinarily rapid pace. As a result, China manufactures almost half of all the cement in the world and the share of industry of all kinds in its economy is a remarkable 44% (UK 20%, US 21%, Germany 31%).

[iv] International Energy Agency, *Key World Energy Statistics*, 2014.

As economies develop, technically sophisticated light industry and services come to play an ever increasing role. Largely service based economies, such as the UK, tend to have lower energy intensity than other developed countries with larger manufacturing or mining sectors, such as Germany, Canada, or Australia, but the difference is not nearly as great as it is from developing countries. Some of the poorest countries, for instance in sub-Saharan Africa, have the highest energy intensities because of the basic nature of their industries and the lack of investment in advanced infrastructure. For instance, power may come from diesel generators and transport may be by truck on poor roads. That is why high international energy prices hit these countries the hardest.

Some of these comparisons might be thought "unfair" because, for instance, a main reason why China has high energy intensity is because it is supplying manufactured goods to more developed nations, such as the UK, on such a huge-scale. But this only accounts for a part of the difference.

The trends are apparent from the most recently published energy statistics for the UK.[v] Between 1970 and 2013 the UK's GDP nearly tripled, but primary energy consumption was essentially static. So overall energy intensity fell by two thirds. The components of that are reasonably clear. Industry was responsible for 40% of final energy consumption in 1970, but this had fallen to 16% by 2013. This was partly because the share of industry in the economy declined. But it was also because the energy intensity of industry declined by 70% reflecting efficiency gains but, also, a shift from basic heavy industry to higher value, higher technology activities.

In this process, however, the UK has "exported" some of its energy consumption as goods once manufactured domestically are now imported from China or other developing nations. While the share of industrial energy use declined the shares of domestic use and transport both increased (from 24% to 29% and from 18% to 36%). There have been big gains in the efficiency of home heating and appliances but, against that, there has been a proliferation of new kinds of appliances and average

---

[v] UK Department of Energy and Climate Change, *Energy Consumption in the UK*, 31 July 2014.

households have fewer people, meaning that more homes are required. Similarly, while vehicle efficiency has increased, so has car ownership and, over most of the period, vehicle weight and power.

It is hard to underestimate the significance of these trends from the point of view of climate change. The greenhouse gas emissions of China, the world's largest emitter will eventually peak and start to decline. Partly this will be the result of policies to enhance energy efficiency and convert to clean energy technology. But to a large extent it will also be the result of the natural evolution of China's economy as heavy industry and manufacturing begins to play a lesser part. This natural trend in the way that developing economies evolve is one of the most powerful forces that is on our side as we work to tackle climate change. Without it we would be in much worse trouble than we are today.

Against this, there are other regions with large populations, in India, Africa, and South East Asia, which are, hopefully, approaching the most energy-intensive phase of their economic development. As discussed in Chapter 4, it is vital that with the advance of technology and the support of richer countries, they should be enabled to follow a somewhat different path from China.

3. Overcoming the Inertia

There is an important distinction between energy intensity and energy efficiency. Suppose that I decide to holiday at home, instead of flying overseas, but spend the same amount of money. That is likely to reduce the energy intensity of the economy, because air travel uses a lot of energy. But it cannot be said to improve energy efficiency because I am not getting the same product. However, if the airline upgrades its engines to make the same flights with less fuel, that is a saving in energy efficiency because they are delivering the same outcome with less energy.

I am now turning to energy efficiency properly defined in this way. It is not a single thing. It is a huge portfolio of options affecting all aspects of business, government, and our private lives. Most of these options are a nuisance, or at least inconvenient in one way or another. You could almost certainly save money by insulating your home more effectively. You would have to find an operator that you trust, make sure that you are in when they need access, and, depending on the specific job, rearrange your house while they are working. How high is that on your to do list? You would save energy by taking the train instead of your car, but perhaps you would rather not be bothered with railway timetables? Most factories could save energy by modernising and "right sizing" heat and power systems. But management and shareholders are probably more interested in exciting new products and markets. In our hearts we know we should do these things but in day-to-day life, quite pressured for many, we can't be bothered.

Energy efficiency policy is about finding ways to overcome this inertia. There are essentially three options; exhortation, regulation, and economic incentives. Unfortunately, as in so many areas of energy policy, the most effective tend to be the least politically acceptable.

4. Exhortation

Exhortation is fine up to a point and almost all government have indulged in it. In most circumstances it isn't very effective. It undoubtedly works best where there is a widely shared national sense of crisis. For instance, after the Fukushima accident, when Japan closed nuclear power stations that contributed about a third of electricity supply, there was indeed such a sense of crisis and the government was extraordinarily successful in persuading the Japanese people to reduce electricity demand. But exhortation can be politically risky. In liberal democracies people don't necessarily welcome being lectured to by the government. In the UK it's called the "nanny state".

I was working in the office of a UK energy minister when the government was trying to cope with reduced electricity supply that was due to a miners' strike. Unfortunately the public did not entirely side with the government against the miners and when my minister urged people to "clean your teeth in the dark" to save electricity, this didn't go down well. The newspapers published photographs of the minister's own house with all lights blazing and he came in for a lot of criticism. Amongst the many letters that he received from the public, I still treasure the memory of one suggesting a new campaign asking people to go to the lavatory in the dark, "and give the nation something to aim for!" For all that, there is definitely a place for well-judged guidance by the government on energy saving options.

5. Regulation

Regulation can be highly effective in the right place. For instance the regulation of vehicle efficiency has been highly effective in Europe, the US, and Japan, and is starting to have a big impact in China. So has the regulation of appliances, such as refrigerators. Regulation of power stations has been effective in reducing local pollution in many countries. Generally, it is politically easier to regulate businesses than individuals. Of course it only works if the necessary technology is available at reasonable cost. Regulating buildings, especially homes, is more difficult. Measures that increase the cost of new homes or impose additional costs on existing home-owners are obviously not going to be popular and the UK government, in particular, has been cautious in this area.

The drawback of regulation is that it is complex and sometimes contentious. The government has to intervene in many different areas and to make judgements, likely to be contested by industry lobbyists, as to what energy savings are reasonably achievable. Probably, industrial managers understand their own options better than regulatory officials so they may have the upper hand. And of course there have to be costly enforcement agencies.

6. Economic Incentives

Raise the price of energy! It's obvious. Then people have the incentive to use energy more efficiently right across the board. The creativity and inventiveness of innumerable actors in many different sectors and with a vast range of possible technologies is brought to bear on the problem. Economic incentives for energy efficiency are at once the most effective and the least politically acceptable. Governments are extraordinarily reluctant to raise energy prices, perhaps for understandable reasons. Energy prices impact directly on the poor, as well as the middle classes and are highly sensitive in most parts of the world. In an opinion poll in 2013, the UK public gave energy prices as the leading threat to the economy.[vi] This topic overlaps with the topic of putting a price more specifically on carbon emissions, discussed in the previous chapter.

As described in Chapter 7, many countries around the world actually subsidise energy, thereby further reducing the incentives for greater efficiency. According to the IEA, these subsidies amounted to $548 billion worldwide in 2013, about four times the global subsidies for renewable energy. Most of these subsidies are given by energy exporting countries making energy available domestically at prices that are closer to production cost than to international market price. Nevertheless, this is a huge incentive for profligacy. Even in the UK, notwithstanding years of rhetoric and special schemes to promote energy efficiency, domestic oil and gas is taxed at a special concessionary rate that is below the standard level of Value Added Tax (VAT) that applies to most other purchases. Energy subsidies are discussed in more detail in Chapter 7.

In many ways energy efficiency is indeed like an additional fuel. But in one important respect it is different. It does not fit the standard commercial model for energy supply in which you explore for resources, build facilities, sell energy, and make a return. With energy efficiency there is no extra energy to sell, indeed there is a reduction in spending, so this commercial motivation is absent. Some would say that this exposes a basic flaw in the way our society is run — i.e. that there is a systematic bias towards getting and spending. It is certainly a big drawback for energy efficiency and it hugely distorts energy supply. In most circumstances it is fairly easy to show that it is significantly cheaper, and of course much better environmentally, to save the equivalent amount of electricity through efficiency measures than to build an additional power station. But there is no commercial driver for the efficiency option.

There are two other problems that exacerbate this. It is very common for the constructors and owners of buildings not to be the people who will ultimately pay the utility bills. The building is leased or rented, or sold on, and the additional cost of high quality insulation or, for instance, of a superefficient boiler cannot be captured in the price. So the constructor will settle for the basic minimum. Secondly, energy users often have much

---

[vi] Retrieved from https://yougov.co.uk/news/2013/09/25/energy-prices-economic-threat/. Accessed 7 December 2014.

higher discount rates than energy suppliers. A big business, such as a utility, may be willing to invest in reliable energy supply options with a rate of return of 10–15% p.a. But you probably wouldn't invest in a more efficient boiler or home insulation unless you are convinced that you will get your money back in 2 or 3 years. In other words you are looking for a return of 30% plus. So there is a bias in favour of more energy and against energy saving.

These are, of course, well known problems and many efforts have been made, both by industry and government, to get around them. The industry solution is the "energy service company". These companies invest in improving the energy efficiency of the host businesses and receive, in return, a share of the consequent savings in energy costs. There are many successful companies that do this, earning a good return for themselves and their host businesses and making a positive contribution to the environment. But it isn't a perfect solution. The drawbacks are obvious. Energy efficiency investment may be disruptive of the business of the host company and, for instance, "right sizing" of energy equipment may reduce flexibility. More significantly, the savings from increased efficiency are often hard to predict because they depend on factors such as the weather and the future activity of the business, quite apart from the performance of new equipment. So a high level of trust is needed between service company and host, and only projects with very good intrinsic returns are viable.

7. Buildings

The need to provide decent homes for large numbers of poor people is one of the primary drivers of economic development in major developing countries such as China or India. Indeed, largely for this reason, almost half the world's construction is currently taking place in China. This is where the megacities of the future are now being created. China is now producing more than half of the world's cement and, as a result, China's cement industry alone is emitting almost twice as much CO_2 as the entire UK economy. Construction is nearing its peak in eastern China and will probably peak in central and western China within the next 10–20 years. Chinese building standards have been lower than those in the West, although they are improving now. Inspection and enforcement are critical. Anecdotal evidence suggests that many new buildings fall well short of the standards that are set, for instance in the thickness and quality of insulation.[vii]

Over the next decade or so there is a major opportunity to improve the efficiency of China's built infrastructure for the future. If this opportunity is not taken then a high level of energy intensity will be locked in for many years to come and many of these buildings will need to be replaced early, which is also a bad outcome for the environment. This is of course a matter for Chinese authorities but it is one of many areas where engagement between Chinese and Western experts is important.

The UK, in contrast, has an exceptionally large stock of older, generally poorly insulated buildings and is adding only about 1% to its housing stock each year. Upgrading existing housing presents greater difficulties, in many ways, than setting high standards for new buildings. Nevertheless, basic measures to improve insulation, to install quality doors and windows, as well as best practice equipment such as efficient condensing boilers, will radically improve the efficiency of most older homes.

Understandably, in their energy efficiency policies, governments tend to pursue social objectives as well as objectives more strictly related to energy and the environment. The recent development of UK policy provides an interesting example.

8. The UK's Example

A large part of the UK's policy has been, and remains, the placing of an obligation on electricity and gas suppliers to achieve specified energy and emissions savings across their customers. This is known as the Energy Company Obligation (ECO). This approach has the merit that it is the supply companies that have the direct relationship with consumers and who, hopefully, enjoy a measure of trust. The approach also gives the supply companies quite a lot of flexibility as to how they achieve the savings which should encourage cost effective methods to emerge. Amongst the

---

[vii] Hirst, N. 2012, *Buildings and Climate Change,* Chapter 2 of *Design and Management of Sustainable Built Environments*, editor Yao, R. Springer.

biggest energy savers to date have been cavity wall insulation and loft insulation. The companies of course recover their costs by increasing the price of gas and electricity. Assuming that their schemes are successful in reducing demand, this does not necessarily mean an increase in total bills. But consumers, and the media, certainly focus their attention on actual fuel prices and the UK government has, therefore, become quite cautious about the extent of these obligations.

A further scheme, known as the Green Deal, was introduced in 2012. The basic idea was to finance energy efficiency improvements through a surcharge on utility bills. In that way the providers of energy efficiency equipment and services could be confident that they would be paid and, the idea was, therefore to be able to make competitive offers with minimal financing charges. Another feature of the scheme was the so called "golden rule" which said that all projects funded under the scheme must be estimated to produce utility bill savings more than sufficient to pay for the surcharge. In other words customers should experience a net reduction in their bills. However, a combination of inertia on the part of customers and, probably, lack of confidence in the scheme and its providers meant that the uptake was poor. According to the BBC, at one point in June 2013, there had been 38,259 assessments under the scheme but only four households had taken it up.[viii]

9. Power Stations

Efficiency improvements in coal and gas power stations are already making a big impact on global energy efficiency. The most important technologies, the Combined Cycle Gas Turbine, and supercritical and ultra-supercritical coal power stations have been described in Chapter 7. However, the impact of building new and highly efficient coal plant is mixed. India is building a new generation of huge "ultra mega" coal generation complexes. They are all supercritical, which means that they will raise the overall efficiency of the Indian coal fleet. Nevertheless, they will lock in substantial emissions of CO_2 for at least 40 years, to a period when India, along with the rest of the world, should be reducing her emissions

---

[viii] BBC News, *Only Four People Sign up to Flagship Green Deal*, 27 June 2013.

to very low levels. China, similarly, is building its new generation of coal power stations, most of them supercritical. As discussed in Chapter 4, both countries currently have a surplus of generating capacity, due in part to the rapid growth of renewables, so investment in new coal power is likely to be curtailed, at least for the time being.

Even the most efficient coal power stations convert less than 50% of the energy in the coal that they burn into electricity and gas power stations less than 60%. The rest of the energy is wasted. When an air cooled power station is in full operation you can see huge cloud-like plumes rising into the air. Many assume that this is smoke from combustion but in fact it is steam from the cooling towers whose sole function is to get rid of surplus heat, in other words wasted energy, from the back end of the plant.

10. Combined Heat and Power

It would obviously be much better if a useful purpose could be found for this wasted heat, a process known as combined heat and power (CHP). However, a full-scale power station produces a lot of heat. To make use of a significant proportion of it you need an industrial operation with a big heat requirement sited nearby or you need a large centre of population and a huge network of pipes to distribute the heat to individual buildings. Not surprisingly, in the past, large scale CHP has tended to be associated with centrally directed economies, such as those of the Soviet Union — though some of these schemes became very inefficient. In some developing countries, where homes and factories are being built most rapidly, large-scale CHP may be achievable, although rapid development is usually the priority, rather than energy efficiency.

In the UK, a mature economy without much growth in heavy industry and with a very slow replacement of the historic building stock, it can be very difficult to promote CHP. In the 1990s, concerned about the impact of the "dash for gas" on the coal industry, the UK government imposed a moratorium on new gas-fired power stations. An exception was made in cases where it could be shown that there was a use for the waste heat, but there were few takers. A number of projects offered to create a business park with access to low-cost steam, but there were few businesses with steam needs on the right scale willing to build or relocate. A number of ingenious uses have been found for waste power station heat, for instance to heat greenhouses, but they are seldom on a scale to use a large share of the heat available.

The most common form of CHP in the West arises when the managers of major industrial plant see the opportunity to build their own power plant adapted to supply their needs for both steam and electricity. Smaller power plants are generally not such efficient electricity generators as larger plant, but having a use for the steam as well makes them economically attractive. Such generators usually operate to meet the day-to-day needs of the industrial plant that they serve, however with suitable incentives they can also contribute to the flexibility of national power networks.

11. Industry

Both India and China have hugely ambitious schemes for improving the energy efficiency of their power generation and industry.[ix] The Chinese scheme, originally the 1,000 Energy Consuming Enterprises Scheme has now been expanded to cover 10,000 Enterprises.[x] It covers power stations, factories, and even large buildings that together account for 70% of the country's energy consumption. This policy plays a large part in China's efforts to meet the voluntary target that China has set itself to reduce carbon emissions by 40–45% in 2020 compared to 2005.

The programme is managed by local governments who set targets for each enterprise. However, central government sets overall targets for the local authorities and officials whose regions fall short are marked down for promotion opportunities. Similarly, enterprises that fail to meet efficiency targets lose out on efficiency awards and have to conduct mandatory energy audits followed by necessary efficiency investments. The

---

[ix] Hirst, N. 2012. *An Assessment of India's 2020 Carbon Intensity Target*, Grantham Institute for Climate Change Report GR4 and *An Assessment of China's Carbon Intensity Target*, Grantham Institute for Climate Change Report GR3.
[x] IPEEC Blog, 15 January 2015, China is on track to meet its 12th Five Year Plan Energy Intensity Target.

scheme has already been credited with saving energy equivalent to 150 m tonnes of coal. As with most other economies, the really big energy consumers are power generation, steel, cement, and chemicals, including oil refining. Because China is developing its infrastructure so rapidly, these efficiency programmes are of immense global significance.

The equivalent scheme in India is called Perform, Achieve, and Trade (PAT).[xi] This covers nearly 500 of the largest manufacturing and power plants. Each plant is given its target after an energy audit. However, unlike the Chinese scheme, there is a trading mechanism for plants to sell any excess savings that they make beyond the target or buy savings if they fall short. In principle this should help to ensure that it is the most cost-efficient opportunities for energy saving that are exploited. The scheme is reported to have saved 31 million tonnes of CO_2 equivalent by 2015 in its first phase.[xii] It has contributed to India's target of achieving a 20–25% reduction in national carbon intensity by 2020, compared to 2005. India's target may seem more modest than China's but it is important to realise that India, as a less highly industrialised society, starts from a much lower level of carbon intensity.

Europe's Emissions Trading Scheme (EU ETS) has been described in the last Chapter. It covers 10,000 of Europe's most energy-intensive power stations and factories. Although targeted on carbon dioxide emissions it also aims to improve energy efficiency. All plant must have quotas equivalent to their emissions. The scheme has been dogged by an excess of quotas, leading to very low prices and, therefore, limited incentive to improve efficiency and reduce carbon emissions. This excess is partly due to the recession that started in 2008 and partly due to political infighting in which each EU member state has sought to limit the burden on its industries. The scheme has also been criticised because of the decision to issue the base level of quotas free to most industries. In future, the intention is that only those industries that face international competition should have free quotas.

[xi]Neelam, S. 2013. Creating Market Support for Energy Efficiency: India's Perform, Achieve and Trade scheme, Climate and Development Knowledge Network.

[xii]De Ashok Kuman, *Overview and Status of PAT Scheme*, Presentation to a Workshop on the PAT Cycle, New Delhi, 13 June 2016.

12. Vehicles

Vehicle efficiency is without doubt one of the most effective ways of reducing carbon emissions. Countries representing 80% of the world's vehicle sales have schemes in place, including the EU, the US, China, and India. Efficiency has improved substantially in all major countries over the past decade and there are ambitious targets for the future. The main savings will come from improvements in the efficiency of the internal combustion engine, from weight reduction, and, increasingly, from hybridisation to incorporate electric power. By 2025, the US and EU vehicle fleets are expected to be about twice as efficient as they were in 2000, a remarkable achievement. When one considers that fully one third of all CO_2 emissions are from combustion of oil, two thirds of which is used for transport, it is easy to see what a huge contribution a doubling of vehicle efficiency can make.

One of the most successful, and best known, schemes for improving the efficiency of vehicles and appliances is Japan's "top runner" programme, which has been operating since 1998. The essential idea is to take the most efficient product in each class and then require manufacturers to bring the average performance of their products up to this level within a certain deadline. One obvious advantage of this approach is that the "top runner" provides an objective and clearly feasible benchmark, limiting the scope for lobbying. In practice, the targets have also been informed by the advice of expert technical advisory committees. Partly as a result of this scheme, the average efficiency of room air conditioners in Japan improved by 68% between 1997 and 2004 and there were efficiency improvements of 55% for refrigerators, 99% for computers, and 23% for cars over similar periods.

13. Conclusion

Most commentators agree that energy efficiency is the biggest player in the effort to avert the threat of climate change. The most effective measures are strong regulation and economic incentives, such as putting a price on carbon emissions, a topic discussed in the last Chapter. Such measures may not be popular with industry or the public. Therein is the dilemma. If we are to address climate change successfully, governments are going to have to find acceptable ways to get tough.

Chapter 9

Markets and Competition

1. Introduction

Energy markets should be as free and transparent, and accessible to both producers and consumers as possible. In that way consumers can access the lowest cost sources of supply and the right signals are sent to investors to build the capacity that is needed for the future.

In practice there are many policy issues, some legitimate and some not so legitimate, that tend to complicate this simple principle, as well as more technical questions of market structure. Governments may have legitimate fears as to whether markets will deliver in their particular circumstances. They may have concerns about energy security and market volatility. And they may have social, environmental, or industrial objectives that they think are more important than the absolute lowest price.

It is generally accepted that the physical transmission and local distribution of gas and electricity are "natural monopolies". It doesn't make economic or social sense to have different companies putting down competing networks of pipes and wires. There are situations where private companies are allowed to profit from building new "merchant" links. For instance the Estlink connection between Estonia and Finland, built in 2006, was financed in this way. But generally speaking the physical networks are regulated monopolies with their prices controlled by government agencies. Especially with modern information technology, this is quite consistent with competitive markets in the production and sale of gas and electricity.

In developing countries there is often a high degree of government control over the whole of the power sector. For instance, the Chinese power sector is dominated by five big state owned utilities and the central government sets targets for investment in the various generating technologies and in the power grid. India also has large government owned power companies and the government in New Delhi sets specific renewables targets for each state, although these targets have not always been met. Not all developing countries have the technical and administrative strengths to run efficient power markets, and capital markets may not be deep enough to finance the needed investment without government help. China has now reached a stage of development where more competitive markets may make sense, and wide-ranging reforms are under consideration, intended to give greater scope to competitive markets.

In the West, although utilities have usually been privately owned, and most governments have sought to stimulate competition in power supply, there is, nevertheless, extensive government regulation and the need to address the externality of greenhouse gas emissions has led to a high degree of intervention in the choice of power station technology.

Thus, in most of the world a high-level of government intervention and control of energy markets is the rule rather than the exception.

2. UK Experience

During the 1950s and 1960s, virtually the whole of the UK energy sector was a government monopoly. Electricity was generated and supplied by the Central Electricity Generating Board (CEGB) which also owned and operated the grid. The British Gas Corporation occupied a similar position with regard to gas. The National Coal Board supplied coal and the Atomic Energy Authority built and operated nuclear power stations.

Was this because governments were so ignorant that they didn't understand the benefits of competition? Not necessarily. The circumstances were different from today's. Many of the companies had been private before the Second World War and they had suffered chronic under-investment. There had been bitter labour disputes. During the war, a very large share of the economy had been devoted to military activities under government management. After the war there was a tremendous shortage of power, and private sector institutions were generally weak and lacking in capital. The government wanted an economy that would be "fit for the heroes" who had fought in the war and who would now be the working men in energy and other industries. The government also wanted to pursue the "white heat" of new technologies developed during the war. In the field of energy, this meant nuclear power, which had special safety and security requirements and which, in the beginning, was closely connected with the development of nuclear weapons.

So it seemed the right thing to do to "nationalise" the energy industries, so that the government could provide the necessary investment and coordination and ensure decent employment standards for the workers. It was also a period when there were big efficiency gains to be had from applying best available technology, in the form of ever larger scale electricity generators, and this did not depend on competition. As natural gas became available from the North Sea there was a need to switch the nation's cookers and boilers from "town gas" to chemically somewhat different "natural gas". This compulsory process was arguably best handled by a government agency.

By the 1970s, however, the inefficiencies and rigidities of this system were beginning to emerge. In the huge monopolistic bodies that had been created the trades unions acquired increasing power. This affected not only wages and conditions but also industrial structure. The UK had been mining coal since the industrial revolution and remaining reserves were increasingly deep and costly to access. Cheaper and more accessible coal was available in many parts of the world. Gas had become more competitive with the arrival of the highly efficient "combined cycle" turbine. But the closure of a coal mine was a politically toxic event and successive governments poured subsidies and investment into the industry. Trades unions and lobbyists for the coal industry, the electricity supply industry, and the manufacturers of boilers, turbines and generators had significant influence on investment decisions. In justifying the costs, politicians of all parties clung to unrealistic myths about the efficiency and prospects of coal mining in the UK. A succession of strikes by power sector unions was a cause of major national and political disruption.

It was Margaret Thatcher's governments of the 1980s, and of her successor John Major, that privatised the energy industries. Government monopolies in electricity, gas, and coal mining were broken up and sold and new competitive markets were established. At one level this was seen, correctly, as an attack on the unions. The struggle with the mining unions and the abrupt closure of large parts of the mining industry were traumatic political and social events that have left a scar. The UK is not the only country where the decline of coal mining raised social issues. We see, today, the difficulties that China faces as national efforts to moderate and then shrink the coalmining industry lead to growing unemployment in coal mining regions. Discontent in coal mining regions of the US has been credited as one of the factors propelling the anti-establishment candidate, Donald Trump, to the presidency.

In many respects the circumstances were favourable for energy privatisation in the UK in the 1980s. The competitive price of gas and the surplus of existing generating capacity meant that considerable price reductions were possible. Contrary to alarms that were raised by some, supply from the privatised industries proved to be highly reliable. Advances in information technology had made it possible to run sophisticated electricity and gas markets in "real time". In other words, trading can take place very close to the time when the electricity is transmitted. Politically, it is much easier to reform the power sector when there is a prospect that consumers will benefit from falling prices. Whether another political objective was achieved, the promotion of popular share-holding, is beyond the scope of this book.

With this perceived success, the UK became an enthusiastic promoter of privatisation and competition in the power sector elsewhere in the world, with mixed results. In retrospect it is clear that some of the countries to which the UK tried to sell energy privatisation, such as Russia and states in the former Soviet Union, were not well prepared, technically or socially, for such radical change. In some regions it was relatively successful. But in other, politically less mature countries, it opened the door for corruption with political insiders getting hold of assets at well below their true value. For any country considering privatisation, this question of whether it can be handled without substantial losses through this kind of corruption is an important consideration. Even in the UK, where in all my personal involvement in five separate privatisations I did not witness corrupt dealing, the government was, of course, well aware that the process of transferring these huge enterprises to the private sector would be profitable for City of London financial institutions with which they were politically allied.

The benefits of privatisation depend, to a large extent, on the creation of a genuinely competitive market. In the UK the operation of the high voltage electricity grid and the gas trunk lines, as well as the physical local electricity and gas distribution networks are still regulated monopolies, although privately owned. However, the acquiring and selling of gas, as well as electricity generation and supply, are subject to competition and there is no price regulation. How competitive are these markets? Critics point out that there are only six major players and a high degree of vertical integration between electricity generation and supply. Prices depend to a certain extent on international fuel prices, especially for gas, and also, to an increasing extent, on obligations placed on the companies by the government to improve carbon efficiency, address fuel poverty, and reduce carbon emissions.

We do not look for perfect competition in other areas of the economy and in other sectors the existence of six major competitors might be regarded as perfectly adequate. However, electricity and gas (and indeed water) may be regarded as a special case. It is not as easy to switch supplier as it is, for instance, to choose between consumer products on a supermarket shelf. Or at least the public tend to believe that it is not so easy and may distrust the process. The supply companies trade, to some extent, on inertia, and governments and regulators have struggled, over a period of decades, to simplify the process and convince the public that it is worth their while to shop around. A major difficulty is that ordinary people are just not interested. Comparing the merits of different electricity or gas suppliers seems like a rather obscure and technical business and many consumers can't be bothered. In 2015 12% of the UK's domestic electricity consumers and 13% of gas consumers switched suppliers, so the market is hardly static. Today, in 2017, the practice of the utilities of charging higher prices to customers who fail to switch is becoming politically intolerable, because these are often the poorest and most disadvantaged. Even a Conservative government, which would normally be committed to competition, is proposing price controls.

Now it is the UK government itself that is undermining competition in electricity markets, at least as far as investment in new capacity is concerned. The wheel has come full circle! Government intervention in the market is like walking into quick sands. Once the first step is taken it is very difficult to draw back. In the UK, the effectiveness of the market was increasingly at risk once the government decided to guarantee the price of electricity from renewables. This might not have mattered very much as long as renewables were a very small share of the market. But now, in 2017, renewables are contributing around 25% of electricity supply in the UK, and a much larger share at times of low demand when the wind is blowing. Other sources cannot compete because renewables continue to run even if their price in the market is zero or even negative. Who would invest in new conventional generation in such a market?

The consequence has been the drying up of investment in conventional power. However, wind generators cannot supply electricity on still days and solar energy contributes no power at night. This has caused growing concerns about security of supply. As a result, the UK government is now buying back-up generating capacity through a series of auctions. The most efficient, and lowest carbon, source of back-up would be new Combined Cycle gas turbines. But these require substantial new investment that remains difficult to justify on the basis of back-up contracts. So far, in 2017, the auctions have mainly attracted diesel generation, and less efficient "once through" gas generators that do not have a combined cycle. They have also financed the keeping open of coal power stations that, on climate change grounds, the government would otherwise have liked to close. Something very similar is happening in Germany, as I have described in more detail in Chapter 6.

Lobbyists are now calling for stricter environmental criteria to be applied to the back-up capacity that qualifies for support. That might solve the environmental problem but it would also increase the cost of back-up.

There is a circular element in this situation. As soon as the government announces its willingness to buy back-up capacity, there is every incentive for the industry to withhold investment that does not receive this support. In effect, almost all new capacity will now be based on government decisions. The arrangement appears wide open for manipulation by the power industry.

This is a mess. It is to be hoped that measures being taken to improve the flexibility of the energy market, including improved demand side flexibility and energy storage, will eventually reduce the need for back-up supply. But this is a long way off, and the jury is still out on how far it can go.

It is arguable that it would now make sense for the UK government to rationalise its *de facto* control of investment in power stations. In other words, to bite the bullet and regulate this investment in much the same way that it already regulates the electric grid, taking decisions on new capacity and setting the rate of return and electricity prices. The market is not contributing much to investment decisions in the UK today.

I would prefer to see a gradual return to competition. This could be achieved by moving to carbon pricing or another system where new renewables were supported on the basis of the value of carbon emissions saved, rather than through price guarantee. This will take time because the price guarantee on existing renewables has been locked in for 15 years. But it could gradually restore a degree of meaningful competition and rationality to the UK power sector.

3. US Experience

The standard model for electricity governance in the US is the Public Utilities Regulatory Commission (PURC). The PURC usually covers a single State and has an elected board. In the traditional structure each PURC has a small number of privately owned power companies that are subject to its regulation and which have a duty to supply consumers in the state. In return, each utility has the right to earn a regulated rate of return on its "used and useful" assets and a monopoly in its area of supply. The term "used and useful" is critical. In order to qualify to receive the regulated rate of return on the cost of a new power station, a utility has to demonstrate not only that it has been successfully commissioned but also that its power is making a positive economic contribution. Thus if an additional power station is needed the utility puts forward a specific project for approval. Provided that the station is successfully brought into operation and that its power does indeed turn out to be needed, its capital costs (including interest during construction) are allowed to enter the "rate base". Electricity prices are regulated so that the utility is just able to earn the regulated rate of return on all its "rate base" investment. A powerful moderating factor in this system is the bond rating of the utility. If the PURC is too aggressive, or "populist", in its dealings with the utility, not allowing it reasonable assurance of a fair rate of return, the bond rating agencies are likely to downgrade the utilities' debt. This has the result that the utility has to pay higher interest rates. This is taken as a highly public signal of mismanagement on the part of the PURC, since everyone knows that these higher interest costs will ultimately fall at the door of electricity consumers.

Variants of this PURC system have operated satisfactorily in many US states for many years. At its best it provides a framework that is democratically answerable, yet provides a stable basis for large-scale investment in energy supply. It was well adapted for reaping the benefits from ever larger generating units and other engineering improvements, which for many years was perceived as the main objective. Prices declined, the rewards to the companies were decent, and everyone was happy.

However, the PURC system also has weaknesses which, eventually, were severely exposed in the US. The inherent problem is lack of competition. Inevitably the utilities themselves have greater depth of expert knowledge than the PURCs and this can lead to PURCs being "captured" by the utilities or at least there has been at times a widespread suspicion that this may be the case. So long as economies of scale continued to deliver price reductions the problem was masked. But this era came to an end as the size of individual generators stabilised. Also, power and gas supply became more complex. As long as the main issues were the building and effective operation of conventional power stations, the potential for distrust was fairly limited. There are plenty of benchmarks for that. But the landscape started to change, fundamentally, from the 1970s.

Dissatisfaction came from a number of sources. In the 1970s, there were gas shortages and (OPEC) delivered a sharp increase in the price of oil. It's easy to forget that in the 1970s about 20% of US electricity came from oil, about the same as gas. Nuclear power brought controversy and in some cases heavy cost overruns. Environmentalists challenged new power and transmission proposals and new environmental regulation increased the cost of generation, especially from coal power stations. Prices were rising and, from the point of view of the elected boards of the PURCs, the politics of simply passing through the costs of the utilities became much more difficult.

In Washington in the late 1980s, a local utility told me that their greatest secret was the functional directory of their business. This was the document listing the names and duties of their staff. It was confidential because relations with their PURC had become so litigious. If the PURC knew who was responsible for what, they could summon them by name for interrogation. Such were the tensions that had arisen between the industry and its regulators. Mutual trust was in short supply.

From the 1970s the US began a transformation of energy regulation that largely mirrored what was happening in the UK. As described in Chapter 2, the Federal Energy Regulatory Commission (FERC) responded to consumer pressure for access to the cheap gas by adopting a range of measures to open up inter-state trade. They also facilitated inter-state trading in electricity, or "wheeling" as it is known in the US. By the 1980s, there was an excess of gas supply, first perceived as a short-term "bubble", which then extended into a "sausage". Many utilities, however, had long term gas supply contracts which effectively locked in higher gas prices. Consumers wanted access to the cheap gas, and the utilities and the regulatory system that sustained their monopolies were seen as the barrier. Energy trading became sexy. This was the era that saw the rise of ENRON, the master energy trading company which rose to be the seventh largest company in the Fortune 500 but collapsed spectacularly in 2001, at that time the largest bankruptcy ever in the US.

To varying degrees the PURCs started to reform, with the intention of making their electricity markets more competitive. The best known, most extreme, and indeed disastrous, of these reforms, was in California. I have already described the sequence of events, in Chapter 3 on energy security. There was a series of rolling blackouts in 2000 and 2001, and in a desperate effort to recover supply, state authorities locked in high energy prices for many years. The Governor lost his job.

There have been many studies of the causes of the California crisis which were, indeed, complex. But the fundamental problem was very simple and, one would have thought, obvious. Prices are the signal that the free market uses to ensure adequate investment and, eventually, supply. If governments regulate prices they cannot then rely on the market to deliver security of supply.

The California crisis brought the reform, or "deregulation", of US energy markets to a shuddering halt. Although some states have competitive markets there are many others who remain content with PURC system described above, in which state regulators control investment and pricing in local energy markets.

US support for renewables has not undermined energy markets to the same extent that it has in the UK or in Europe because it is not based on price guarantees. The main incentives in the US are tax credits, which amount to federal government contributions to the capital cost of new renewables, or portfolio standards, usually set at state level, in which all market participants are required to contribute a fixed percentage of renewable supply.

4. Can Markets Deliver the Investment that is Needed?

Competitive electricity markets should be able to provide the necessary incentives to ensure adequate investment in generation. That was the UK's experience during the period from the late 1980s to the early 21st century, until the advent of special incentives for renewables started to distort the system. In a state of perfect competition the incentives to invest may be rather low. But perfect competition rarely exists in real markets. The major incumbents in electricity supply generally have some degree of market power, even when the market is regarded as competitive. That is no different from other sectors of the economy. Also, the technology of electricity generation is constantly improving. It is worth investing in new plant because it is more cost-efficient than the old, and not just where there is a shortage of capacity. Older, less efficient plant can be available to run at peak times.

Even with a competitive market it makes sense to have a competent system operator regularly reviewing the prospective adequacy of generating capacity 4 or 5 years into the future. That can provide useful information to the industry and it could provide a safety net in the unlikely event that a serious shortage was developing. But this should not be an alternative to the forces of demand and supply as the regular means of securing adequate investment.

5. Benefits and Perils of Government Intervention

The highly competitive US gas market, described in Chapter 2, is an excellent example of the benefits of a fully functioning energy market. It has led to the development of new gas production technology, jobs, lower costs to industry and domestic consumers, and now a thriving export market. The two decades of relatively free electricity markets in the UK were also successful in economic terms. Besides its own intrinsic efficiency, the free market approach has the great merit of taking economic decisions out of the hands of politicians. When politicians take decisions on energy investment there is always a risk that they will fall prey to lobbyists and to politically convenient myths.

In the 1970s and 1980s, the British government poured money into advanced nuclear designs, such as the fast reactor and the, now forgotten, Steam Generating Heavy Water Reactor. Almost all of this was wasted. Perhaps some of these technologies will eventually find a place under the sun, but the timing was wrong. I am afraid that the same may be true of the international effort to advance nuclear fusion.

Government intervention does not have to be anti-market. Indeed, the market can provide the most efficient way to achieve policy aims, including climate objectives. Putting a price on carbon, through taxation or other means, can be a highly efficient way of achieving reductions in carbon emissions. I discussed this further in Chapter 6. However, for as long as governments fail to put a price on carbon emissions, there will always be a conflict between the market and the climate.

Even with a price on carbon, there may still be a need for governments to intervene to bring new technologies forward. This has always been fairly high-risk because one can never be certain of the potential of a new technology until it has been demonstrated. The intervention that Western European governments have made to guarantee the price of renewables has had a high cost, but it has arguably been justified by events. These countries have achieved a remarkable penetration of renewable technologies in a relatively short-time. Even more important, these interventions have driven spectacular reductions in the costs of wind and especially solar power, which have transformed the global outlook for carbon abatement.

For instance, in recent years, Germany has been subsidising renewable energy to the extent of about $25 billion p.a.[i] Between 2010 and 2015 the cost of utility-scale PV fell by two thirds[ii] and the prospects for solar power were transformed. German support, enabling the industry to expand on a huge scale, undoubtedly played an important part in this. The outcome was not so successful for German industry, because, as a part of this process, manufacturing of solar units largely shifted to China.

The impact on electricity markets from these interventions has been severe, and governments are only now trying to pick up the pieces. Now that wind and PV solar power are mature and competitive technologies, governments should move, in a measured way, to more market friendly measures. Renewables should be subsidised according to the value of emissions savings, and their price should be set by the market. One of the big advantages of this is that the renewables industry itself will have the incentives to contribute to the challenge of managing variability.

The UK now faces the problem of how to introduce new nuclear generation, a technology which may be needed to meet climate targets. Nuclear power is definitely not a new technology, but building a "first of kind" new nuclear power station, after 20 years in which the domestic nuclear power industry has been allowed to run down, poses similar problems. The first station, Hinkley Point C, is being supported by price guarantees. This is looking expensive, and it might be less costly, as well as more market friendly, for the government to make a direct contribution to the capital costs of the next station.

6. Conclusion

Full competition in domestic electricity markets remains, in global terms, a relative rarity. In the developing world, where administrative, financial, or technical resources may be limited, it makes sense for the government to exercise a high degree of control and leadership. China has emerged

---

[i] Bloomberg View, *Germany's Green Energy is an Expensive Success*, 22 September 2014.
[ii] International Energy Agency, *Energy Technology Perspectives*, 2016.

from this phase and is now considering wide ranging energy reforms intended to increase the role of markets. Hopefully, other developing countries will follow in due course.

In the West, the market has generally been allowed to play a greater role. The US has the most competitive gas market in the world, described in detail in Chapter 2, and the economic benefits have been considerable. However, efforts at reform in the US electricity market have been slowed since the disastrous outcome of misguided reforms in California.

In the UK, competition in electricity and gas markets flourished from the privatisation of the industries in the 1980s until the early 2010s. But vigorous intervention to guarantee the price of renewables, and the consequent need to introduce capacity payments for conventional power, has since undermined the ability of the market to incentivise new investment.

Free markets are the most efficient means of incentivising investment in new capacity, including low-carbon capacity. With reasonable oversight they can provide a high-level of energy security. However, they are not compatible with the regulation of price. Also, at an early stage in the introduction of new technology some support outside the market is probably desirable.

There is a strong political element in energy regulation. Some governments will probably always want to regulate energy prices and, therefore, will have to manage new investment. Other governments can benefit from a more competitive approach. One size does not fit all. The UK's effort to sell its privatisation model to countries at a different stage in their economic development was misguided.

The UK has got itself into a muddle, and I hope that over time, a more market-based approach to incentivising low-carbon capacity will be introduced. The US is in many respects an admirable example of competitive energy markets, but the misconceived and disastrous reforms in California have cast a long shadow over further reform.

The genuinely competitive electricity market is a delicate flower that does not flourish in all soils. Governments have many reasons to want to interfere, some of them quite legitimate. Nevertheless, decision-making by governments is sub-optimal and may be very costly. Where there is genuine potential for fully functioning competition, governments should ensure that their interventions are as market friendly as possible.

Chapter 10

Global Energy Governance

1. Introduction: Why Does Global Governance Matter?

This chapter is about the institutions for international energy cooperation between governments, and how they need to adapt to meet today's energy policy challenges. I review the main organisations as they exist today[i] and then I consider the need for reform and how this might be achieved.

One might think that the structure of international institutions was rather a second-order issue. Surely where there is a will to address problems collectively, appropriate organisations will spring into existence? But in fact, this is far from the case. It is not an easy thing to change international bodies. Their membership includes many nations, each with their own interests to pursue. Once basic roles and functions are opened for review there is a chorus of different demands and it becomes extraordinarily difficult to get the toothpaste back into the tube. Reform

---

[i]The author has been working on these topics since December 2012, firstly with Chatham House, and then in a joint project with China's Energy Research Institute. Outputs are available on the website of the Grantham Institute, Imperial College. They include; *"The Reform of Global Energy Governance"*, Grantham Institute Discussion Paper No. 3, December 2012; four joint reports of the Grantham Institute Imperial College and the Energy Research Institute of China's NDRC on Global Energy Governance Reform and China's Participation, November 2014, November 2015, March 2016, and (final report) June 2016.

"consumes huge amounts of political energy"[ii] and tends to be undertaken only at times of major crisis. Meanwhile the shape of existing institutions has an important influence, for good or ill, on international policymaking. You need well-managed organisations and effective secretariats to formulate, coordinate, and carry out plans and programmes on the scale required. It is the agenda of strong international organisations that define world priorities for international action and focus the collective efforts of humankind.

We should not view the reform of global energy institutions in isolation from other geopolitical issues. Sharing in world leadership, in general, is an essential aspect of the peaceful rise of emerging nations. Energy is only one of the areas in which global governance is having to evolve. In finance, trade, and other areas, also, established institutions are having to adapt to the rise of countries such as India and China. The International Monetary Fund (IMF), the World Bank, and the World Trade Organisation (WTO) are good examples. The process of adjustment has been patchy and in some cases sclerotic. China's recent decision to set up a new, separate, international financial organisation, the Asian Infrastructure Investment Bank (AIIB), is a symptom of that.

The peaceful rise of emerging nations will depend, amongst other things, on the successful adaptation of governance bodies, so that they can contribute to world leadership on important topics in a way that is commensurate with their new stature. Energy is no exception to this. In fact, energy could provide a leading example because there is such a high degree of common interest in solving world energy problems.

This does not imply that global multilateral cooperation amongst governments is the be-all and end-all of energy policy. Bilateral cooperation is also important. The joint announcement on climate change by the US and China was a turning point in the lead-up to the 2015 Paris climate summit. They gave strong backing to the process as well as announcing their own national contributions. There are also significant regional collaborations on energy. And of course, the main actors in day-to-day energy supply, demand, and investment are profit-making institutions

---

[ii] UK Prime Minister Cameron's report to the G20 on global governance generally, November 2011.

operating within general international frameworks of trade, investment, intellectual property, and disputes resolution. One of the main objectives of multilateral government cooperation is simply to keep these avenues for trade and investment open. But government multilateral organisation is of vital importance in a world where so many of our energy objectives are global in nature and where cooperation between developed and developing nations is of the essence.

2. Today's International Energy Institutions

There are many international bodies and initiatives concerned with different aspects of energy policy and technology. It's like gazing at the stars. The more powerful your telescope the more you will see, especially if you include bilateral as well as multilateral organisations. That isn't necessarily a bad thing. Energy is a diverse topic and we don't need to be too tidy-minded about different positive contributions. International institutions are notoriously difficult to kill off, so no doubt there are those that have passed their sell-by date. But there are a lot of bodies doing valuable, often highly specialised, work. In what follows I will be concerned only with major bodies with some pretensions to influence government energy policies, and enhance cooperation, at a high-level, relevant to the global stage.

Some of the bodies listed below are energy specialists. But of course, there are many non-energy bodies and agreements that have very significant impacts on energy. These include the G20 and the G7. The UN, of course, also has a much wider remit than energy. Its Framework Convention on Climate Change (UNFCCC) is also not concerned exclusively with energy although much of its application is in the energy field. Later in this section, we will come to the WTO and the World Bank. They too are not energy specialists, but their influence on world trade and finance is highly significant for energy.

2.1. *The United Nations*

Several organisations of the UN, for instance, those concerned with economic and industrial development and poverty alleviation, have some interest in energy. These include the UN Development Programme (UNDP), the

Food and Agriculture Organisation (FAO), the UN Industrial Development Organisation (UNIDO) and the UN Environment Programme.

This is not the place to review the performance of the UN. At its best it can mobilise world action to address crying needs. At its worst it is a cockpit for political point-scoring. It inevitably reflects our politically fractured world.

In 2015, the UN adopted 17 Sustainable Development Goals intended to end poverty, protect the planet, and ensure prosperity for all.[iii] Two are highly relevant to energy. These are goal number 7, "Ensure access to affordable reliable sustainable modern energy for all" and goal number 13, "take urgent action to combat climate change and its impacts". Former UN Secretary General Ban Ki-Moon launched a Sustainable Energy For All initiative in 2011. Most of this book is about how we could meet these goals and the unavoidable tensions that exist between them. The goals are important because they provide a sense of direction. They have been described as a "vision to mobilise action" and virtually all world governments are formally committed to them. They may be idealistic but they point in the right direction. They are no more than visions until governments and others act in practical ways to bring them about.

The area of energy policy in which the UN has been most active is, of course, climate change, through the United Nations Framework Convention on Climate Change (UNFCCC). The achievements of the UNFCCC were described in Chapter 5. They are considerable. The agreement reached in Paris in 2015 was an important step forward in world climate politics. President Trump's recent announcement, in June 2017, that the US will withdraw from the UNFCCC is a serious setback, of which the full implications are hard to judge at this point. However, most of the rest of the world is committed to stick with the agreement, so it will certainly still be influential and important.

The UNFCCC has created a Green Climate Fund and a Technology Mechanism, both of which could make important contributions to help developing nations to frame and deliver sustainable energy policies. The Green Climate fund is intended to be one of the main financial instruments through which the rich nations help poorer nations to develop and implement

---

[iii] Resolution of the UN General Assembly on 25 September 2015; *Transforming our World: the 2030 Agenda for Sustainable Development.*

their low-carbon development strategies. The Technology Mechanism aims to link poorer nations with the sources of expertise around the world that are best able to assist them with technical aspects of their plans. At the moment, in 2017, they are at an early stage of development and far from delivering their ultimate potential.

All major governments have delivered their Intended Nationally Determined Contributions under the UNFCCC, discussed in more detail in Chapter 5. The process of "ratcheting up" these Contributions, within the framework of the Paris agreement, is now at the centre of the global climate mitigation effort. UN institutions will play a leading role in that.

The UN is where the claims of leading nations to world leadership are played out. It has near universal membership and for many nations it is virtually the only theatre in which they can register on the world stage. For instance small island states threatened by rising sea levels are represented only at the UN. It was pressure from the Alliance of Small Island States (AOSIS) that led to the inclusion in the 2015 Paris Agreement of the aspiration of containing global warming to 1.5°C, even tougher than the agreed target of 2°C.[iv]

The UN has unquestionable legitimacy to lead on global issues such as climate change. But topics raised at the UN are thrown into a highly contested arena and are prone to become mired in wider political debate. It is sometimes easier to make progress in organisations with more limited membership, such as the G20, or operating at a less political and more technical level, such as the International Energy Agency (IEA). The UNFCCC will remain the organisation for top-level climate negotiations but it is not the body for "bottom up" analysis and cooperation on the delivery of national energy policies and strategies needed to achieve global climate targets.

2.2. The G20

One might think it was obvious that there should be a forum for leaders of the world's major nations, such as the G20, but in fact this was slow to evolve. Originally, the main forum for heads of leading nations was the

---

[iv] Press release of the Alliance of Small Island States (AOSIS), 8 June 2015. *Small Island States propose below 1.5°C global goal for Paris Agreement.*

G7, (Canada, France, Germany, Italy, Japan, UK, and US) and eventually the G8, when Russia joined in 2005. This was a club of developed nations. With the rise of emerging economies it became increasingly clear that these countries could not exercise world leadership on their own. From 2005, the G8 invited "plus 5" emerging nations (Brazil, China, India, Mexico, and South Africa) to join them for at least part of their discussions. A rather patronising arrangement, in hindsight, but it did at least widen the debate at the time. The first G8 Plus 5 summit meeting was held under UK chairmanship at Gleneagles in Scotland. For the first time it placed climate change firmly on the G20 agenda. The summit initiated a "dialogue on climate change, clean energy, and sustainable development", as well as a "Clean Power Action Plan".[v] It made a notable contribution on climate issues, working with the IEA as its secretariat.

However, it was the financial collapse of 2008 that exposed the limitations of G8 Plus 5 and led to the creation of the G20. Such was the gravity of the global crisis that it was obvious that it could only be tackled with the full engagement of developing countries, especially China which was already the world's second largest economy. China and other major developing countries had been members of the Plus 5, but it was only with the advent of the G20 that they became equal participants round the table. In addition to the members of G8 Plus 5, the G20 includes Argentina, Australia, Indonesia, South Korea, Saudi Arabia, and Turkey. The European Union also counts as a member, making 20. So those EU countries who are not themselves G20 members also have an indirect say. The G20 had its first meeting at the Pittsburgh summit in 2009 where leaders declared that it was "the premier forum for our international economic cooperation". The G20 not only includes developing as well as developed nations but also from an energy perspective, it includes producers as well as consumers and, in Saudi Arabia, the leading member of OPEC.

As with the original G7, the preparation for G20 summit meetings of heads of government is undertaken by "Sherpas" from each country as well as a range of specialist groups and working parties. The Sherpas are personal representatives of the heads of government, chosen for this purpose. Often they are very senior foreign service officials. The discussions

---

[v] The G8, *The Gleneagles Communique*, 2005.

between Sherpas, which are not always fully reflected in summit *communiqués*, are a highly influential part of the summit process.

The G20, as its origin implies, is primarily an economic and financial body. However, in recent years it has taken a growing interest in energy. China has long favoured an energy role for the G20, as a top-level forum where China is a fully equal partner with the developed countries. At the Brisbane G20 summit, under Australian chairmanship, in 2014, G20 leaders agreed to work together on nine "G20 Principles on Energy Collaboration", summarised as follows:

1. Affordable and reliable energy for all.
2. More representative energy institutions.
3. Open, transparent, and stable energy markets.
4. High quality data and analysis.
5. Cooperation on energy security and emergency response.
6. Phase out inefficient fossil fuel subsidies.
7. Sustainable development, "consistent with our climate activities and commitments" including energy efficiency, renewables, and clean energy.
8. Encourage clean energy technology.
9. Coordination of international energy institutions.

The fact that it was possible to agree on such a list demonstrates how much the energy policies of developed and developing nations, producers and consumers have in common. But the detailed wording also illustrates the tensions. There is strong support for a market approach with a high level of transparency. This reflects the views of the developed economies who belong to the IEA. But "ensure affordable energy for all", rather reflects the needs of developing countries and mirrors the UN Sustainable Development Goal. Markets could no doubt achieve affordable energy for all, but it is questionable whether they could "ensure" it. Similarly, markets, especially if well informed, may become "stable", but this word could also be taken to cover the efforts of the Organization of the petroleum Exporting Countries (OPEC) to manage price levels. Probably the most hard-fought language is in item 7, which reads in full, "Support sustainable growth and development, consistent with our climate activities and commitments, including by promoting cost-effective energy efficiency, renewables and clean energy". The G20 includes EU countries who are committed to limiting climate change and to renewables, but also India, Argentina, Saudi Arabia, and Russia who are more sceptical. India especially gives a much higher priority to economic development than to climate mitigation. The agreed wording includes climate change and renewables, but only in a subordinate clause in an item about "sustainable growth and development". The words "our climate activities and commitments" are there to show that no one is accepting any new climate commitments in this statement. The call for more representative institutions reads in full, "Make international energy institutions more representative and inclusive of emerging and developing nations". The IEA is the only major energy body that specifically excludes developing nations from its membership so it is reasonable to conclude that this is directed, at least in part, to the IEA.

How significant were these agreed "Principles"? That depends on the extent to which future chairs of the G20 rely on them as the framework for their energy initiatives. Turkey and China, the chairs in 2015 and 2016 respectively, have had limited success in building on the Principles, and have largely followed their own agendas.

Energy officials of the G20 now meet regularly as the Energy Sustainability Working Group (ESWG) to review global energy topics and support summit leaders. The progressive maturing of this group is definitely contributing to mutual understanding on global energy topics, and has supported the G20 in launching programmes on energy access, renewable energy, energy efficiency, and climate finance. None of these programmes is of earth shattering impact, but they are pointing in the right direction and will make worthwhile contributions. However, they do not, as yet, live-up to the overarching ambitions of the agreed Principles.

There are also supposed to be regular meetings of G20 Energy Ministers. But so far these meetings have not identified any major topic where they can make a unique contribution, and attendance at Ministerial level has started to fade. The German G20 presidency, in 2017, is not holding a meeting of energy ministers. This is regrettable but no doubt reflects the lack of real progress in previous ministerial meetings.

The G20 is an annual meeting of leaders with a revolving chair and no permanent secretariat. The extent to which energy appears on the agenda depends on the priorities of each year's presidency. Some are more interested than others. There is a troika, of past, present, and future presidencies, intended to maintain some degree of continuity. There have been calls, from China and others, for the G20 to acquire its own permanent energy secretariat, but this has been resisted strongly by the US and Western governments as a whole. They want to preserve the ethos of the G20 as a meeting of leaders to discuss current problems that is not weighed down by institutional structures and agendas. Of course, they may be influenced also by the fact that the West already has its international secretariat in the form of the Organisation for Economic Co-operation and Development (OECD), with which the IEA is associated.

The G20 played a major part in rescuing the world from the financial crisis of 2008, ensuring that China and other major developing nations were participants in urgent efforts to boost international credit. It can be an effective body when there is a critical shared objective. But in other circumstances, its effectiveness is limited by lack of continuity, lack of an effective secretariat, and lack of shared vision. The G20 can be at its most effective when it provides leadership to existing international bodies, using its authority to guide their work, or promotes reforms, as it has done in the field of finance. In most countries, international governance issues are the preserve of the Prime Minister's office, or the foreign service. This means that the Sherpas are a suitable group for discussing governance issues.

The G20 can exert leadership in the field of energy, as demonstrated by the Principles on Energy Collaboration agreed in Brisbane. Continuity is important. The tendency of each new Chair to rewrite the energy agenda undermines credibility. The G20 should stick to the agreed Brisbane priorities as a framework for its energy work. It is also important to follow through with existing initiatives on energy access, energy efficiency, and climate finance.

The G20 can play a vital role in the reform of international energy institutions. A more inclusive IEA could provide the secretariat that the G20 needs to be effective in the field of energy, in much the same way as the IEA has supported the G7 and its predecessors.

2.3. *G7 and BRICS*

Notwithstanding the advent of the G20, the leaders of the developed world continue to meet as the G7 (for the time being these countries meet without Russia). The G7 has, effectively, given way to the G20 as the premier economic forum. The developed nations of the "West" that it represents still accounted for some 47% of the global economy in 2015.[vi] So it remains an influential body. The G7 has its own energy programmes, and it benefits from the full support of the IEA, because all G7 countries are members.

A group of leading emerging economies, the Brazil, Russia, India, China, and South Africa (BRICS), also meet annually at heads of government level. At their March 2012 meeting in New Delhi, the BRICS also decided to explore, "Multilateral energy cooperation within a BRICS framework". So far they have made little progress. The New Development Bank, founded by the BRICS, has, however, started to invest in renewable energy.

2.4. *The International Energy Agency (IEA)*

The IEA is by far the most substantial and influential body for international energy cooperation. It is the only major international body with the capability to engage and analyse energy policy in the round: that is to say, including environmental, supply, and security issues, as well as the range of energy technologies, how they fit together in national energy systems, and policies for deployment. It is concerned with demand as well as supply and, therefore, with energy efficiency — possibly the most important topic of all for addressing the world's energy problems. This broad range is important because energy policy is highly integrated.

The IEA is regarded as primarily a consumer body. Certainly its origins lie in the need to protect consumers from a supply crisis. But it is important to remember that the IEA membership also includes major producing countries such as the US, Norway, Canada, the Netherlands, the UK and Australia. Probably the crucial distinction is that the IEA members are committed to open energy markets in a way that would not be consistent with OPEC membership.

The IEA was founded in the 1970s in the heat of the Arab oil embargo. It was made an associate of the OECD, largely in the interests of speed. Its founding treaty, the International Energy Programme, is largely concerned

---

[vi] Barry, P. Not-so-great expectations: The G-7's waning role in global economic governance, Brookings Institution, May 2016.

with detailed mechanisms for building stocks and sharing oil in emergencies. Because of the original OECD affiliation, the Treaty also confines membership to the developed nations who belong to the OECD. The IEA's 29 members now include most of the countries that belong to the OECD. When the IEA was founded these countries accounted for the great preponderance of international oil demand. Today IEA members account for less than half of world oil demand and this ratio is declining steadily.

The IEA's Governing Board is composed of senior officials of each member country, who meet regularly at the IEA's head quarters in Paris. Every other year, the Governing Board meets at the level of energy ministers. The Governing Board has senior committees that deal with emergency planning, the outlook of energy markets, wider energy policy issues, and cooperation with countries who are not members, such as China, India, Brazil etc.

The IEA has sometimes been caricatured as being primarily concerned with oil security. However, in recent years the IEA has become much more engaged with climate change issues, with the demand side and with alternative energy technologies. It has greatly strengthened its links with major developing countries and is aiming to take this further. If the IEA can achieve its ambition of becoming a genuinely global body then it can play a central role in addressing the world's energy problems. It can provide the energy policy support that the G20 needs in order to provide the necessary top-level leadership. And its original energy security role will be greatly enhanced by the participation of major emerging economies in its emergency planning.

The IEA has seven main functions:

1. It reports on the short and medium-term oil and gas market outlooks.
2. It is the most authoritative source for much international energy data. For instance, the UNFCCC relies heavily on IEA energy data for their assessment of CO_2 emissions.
3. It regularly reports to other international bodies such as the G20 and the Clean Energy Ministerial on global energy developments.
4. It coordinates the oil emergency preparedness of its members and their collective response to supply disruptions. For this purpose, importing member countries are required to hold strategic oil stocks equivalent

to 90 days of imports, a total investment of some $200 billion in oil security.

5. It coordinates energy policies through discussion, analysis, and regular "peer review" of the policies of each of its members.

6. It reviews international energy developments and publishes extensive analysis, including the influential *World Energy Outlook* and *Energy Technology Perspectives*.

7. Through a network of more than 40 technology collaborations it coordinates and promotes the development, demonstration, and deployment of technologies to meet the challenges of the energy sector. This technology network, unlike the IEA itself, is open to developing countries and has genuinely worldwide reach.

The IEA is well aware of the need to adapt to the modern world and is making major efforts to strengthen its strategic role, including the activation of an "Association" with leading emerging economies. China, India, Indonesia, Morocco, Singapore, and Thailand are already members of this Association.

The Association has no legally binding obligations and implementation is quite flexible, so as to reflect the needs of each country. It is essentially an invitation for each member to share in the work of the IEA and contribute to its deliberations in ways that make sense to both parties. It clearly falls short of full membership. It confers no voting rights and, equally, no formal obligation to comply with IEA stock-building requirements or emergency-response rules. Nevertheless, if taken up to the full, the Association could lead to participation in almost all the activities of the IEA. Ministers of the IEA and Association countries described it as a "key step towards building a truly global energy organisation"[vii]. The next steps in this process are discussed later in this chapter.

The Clean Energy Ministerial

The Clean Energy Ministerial (CEM) includes 23 governments of some of the largest energy consuming and producing countries. It does a lot of

---

[vii] *Joint Ministerial Declaration on the Occasion of the 2015 IEA Ministerial meeting expressing the Activation of Association*, Paris, France, 18 November 2015.

work across a wide range of policy and financial issues and it also enhances coordination on policies in areas such as energy efficiency, power system planning, and electric vehicles. It covers a wide and evolving range of technologies and infrastructure issues.

In 2016, the members of the CEM agreed to base its secretariat at the IEA. This was a a positive sign of the progress that the IEA is making towards genuinely global reach as well as a rare and welcome step towards the rationalisation of global governance structures.

2.5. *Technology-Specific Bodies*

From the 1990s, as the threat of climate change became more apparent, world leaders looked for international institutions that could lead the effort to deploy clean technologies and improve energy efficiency. They did not entirely turn to the IEA because it was not sufficiently inclusive and, instead, a new generation of bodies with specific roles was created, more or less overlapping with the work of the IEA. These included the International Renewable Energy Agency (IRENA), the International Partnership on Energy Efficiency Collaboration (IPEEC), the Global Carbon Capture and Storage Institute (GCCSI), the Clean Energy Ministerial, the International Partnership on Hydrogen and Fuel Cells (IPHE), and the Renewable Energy and Energy Efficiency Partnership (REEEP). All these bodies and initiatives play a useful role in relation to their specific technologies but, arguably, there are too many of them with overlapping functions and each competing for the time of energy officials. The absence of a central organisation for energy makes it difficult to coordinate their effort.

This process of proliferation of energy bodies and fragmentation of energy governance has to some extent weakened the IEA in recent years.

Now, hopefully, as the IEA pursues its modernisation, a somewhat greater degree of coherence is returning. The decision of the Clean Energy Ministerial, locate its secretariat at the IEA was a positive step. There are now also proposals to create an energy efficiency hub at the IEA, in cooperation with IPEEC. Other bodies will definitely want to remain separate from the IEA but there is a trend towards closer cooperation. IRENA, for instance, was originally promoted by the German government specifically as a rival to the IEA, which the Germans regarded as not sufficiently committed to renewables and too positive on nuclear power. However, today there is a high-level of cooperation between the IEA and IRENA at a practical level.

2.6. The Organisation of Petroleum Exporting Countries (OPEC)

Although dominated by Middle East oil producers, OPEC also includes producers in Africa and Latin America. The full membership is Algeria, Angola, Ecuador, Iran, Iraq, Kuwait, Libya, Nigeria, Qatar, Saudi Arabia, UAE, and Venezuela. This is a diverse group, and there are some powerful internal tensions, especially between Saudi Arabia and Iran who are rivals for regional dominance. There are also divisions between countries such as Saudi Arabia, which has large financial reserves and is able to play a longer term strategic game, and others, such as Venezuela, who lack such reserves and for whom even short-term declines in oil prices spell economic catastrophe.

OPEC countries control about 40% of world oil production and this share is increasing. This share includes almost all of the really low-cost oil. Saudi Arabia is by far the most significant player because they alone have been willing to keep substantial production capacity in reserve. It is the capacity to hold back oil that could have been produced when prices are considered too low or, conversely, to produce from oil fields previously held in reserve when prices are considered too high, that gives Saudi Arabia a unique potential to influence world oil prices.

OPEC operates as a cartel with the stated aim, to "devise ways and means of ensuring the stabilization of prices in international oil markets with a view to eliminating harmful and unnecessary fluctuations".[viii] As described in Chapter 2, OPEC is certainly not the first in the field to attempt to manage oil prices. There is a long history. OPEC has been successful in influencing the price of oil over extended periods but, like all cartels, it eventually runs out of capacity to buck the underlying market.

[viii] *OPEC Statute* 2012.

This was what happened in 2014, when Saudi Arabia decided not to accept the rapid continuing loss of market share that was consequent on its efforts to keep the price of oil above $100 per barrel. The price collapsed rapidly from $120 to below $50, and then to below $35 by February 2016. In June 2017, it has recovered to just over $50.

Some see Saudi Arabia's decision to let the market rule specifically as an attack on the US shale oil industry. If that was so then the shale industry has proved more resilient than expected. Others have pointed to the strategic struggle with Iran, arguably less able to live with low prices. Perhaps there is also a fear that world climate policies will eventually reduce the value of oil left in the ground. However, the main reason was less complicated. The simple fact was that Saudi Arabia was paying an unacceptable price, in terms of lost market share, for its efforts to manage oil prices. The oil industries of the US and Iran have certainly suffered, and Saudi Arabia may not be too unhappy about that. But probably Saudi Arabia will be willing to stabilise oil prices at a higher level as soon as the changing balance of demand and supply, or the willingness of other oil producers to share the burden, make that possible.

It's easy to see that in recent years OPECs activities have exacerbated price volatility. Has OPEC benefited its members over the long term? That isn't absolutely clear, but probably yes, because it has been able to drive up prices over extended periods, which is especially significant considering the very low prices that prevailed before OPEC came into existence.

The latest position, described in Chapter 2, is that OPEC has achieved an uneasy alliance with Russia and is once again cutting production. This is sustaining an oil price just above $50 per barrel. There is no certainty how long this will last or how effective it will be in mopping-up excess supply in the market. As a general principle, I would expect that OPEC's ability to adjust production levels, albeit marginally, will again have an important impact on oil prices as the market eventually rebalances.

The West has a complex relationship with OPEC. It disapproves of cartels and, generally, prefers lower oil prices. But it also suffers from price volatility and OPECs efforts to manage the most extreme manifestations are, privately, not unwelcome. This was illustrated during the price collapse of 1986. George Bush, then US Vice President, making a tour of the Middle East, said, "I think it's essential that we talk about stability and that we not just have a continued free fall like a parachutist jumping out without a parachute". [ix] Of course George Bush has oil industry connections, and that was not the formal position of the administration. In public the US government can be expected to support the lowest possible oil prices, because that suits US motorists. But in private, as in the example that I have given, there is concern about the damage that sharp price falls can do to the US oil industry.

Although OPEC generally works to keep oil prices at the highest sustainable level, the organisation, and especially Saudi Arabia, recognise that sharp price spikes represent a threat to the market and may provoke further volatility in the future. At times of extreme volatility, therefore, Saudi Arabia is the first port of call. The IEA would not, today, undertake emergency stock release without first consulting informally with OPEC, and especially Saudi Arabia, about the state of the market and their potential and willingness to make more oil available. In 2008, when oil prices peaked at over $140 per barrel, it was Saudi Arabia who called an emergency oil summit in Jeddah, with major consuming as well as producing countries, to consider measures to calm the market. We will return to this later.

2.7. *The International Energy Forum*

The International Energy Forum (IEF), which was set up in 1991, is the major body for dialogue between oil-consuming countries and OPEC members. Saudi Arabia has always take a leading role in the organisation. It has very wide membership including developed and developing countries, producers and consumers. It has a small secretariat, based in Riyadh, and, since 2011, a formal Charter. The Forum "serves as a neutral facilitator of informal, open, informed and continuing global energy dialogue among its membership of energy producing and energy consuming States, including transit States". [x]

Generally speaking, the Forum is a body for exchanging views and creating networks at high level rather than making policy. Its large and diverse membership makes it an unwieldy deliberative body. The Forum's

---

[ix] Learsy, R. *Breaking the Middle East Oil Cartel*, Encounter Books, 2007.
[x] International Energy Forum Charter, Riyadh, 22 February 2011, Section 1.2.

ministerial meetings, held every other year, are major events in the energy calendar, well attended by governments from around the world.

The IEF secretariat has conducted studies on a variety of topics. But its most important role is as a facilitator and coordinator of joint projects and events with OPEC and the IEA. The G20 can commission work by the IEA and OPEC, coordinated by the IEF, that it could not commission from any one organisation because of the difference in membership. This group of three organisations, though it cannot be said to be highly effective, is the nearest thing that we have today to a fully inclusive energy organisation.

Most of the products of oil consumer/producer dialogue have been coordinated by the IEF, including the outcomes of the Jeddah oil summit of 2008. Of these, the most important has been the Joint Organisations Data Initiative (JODI) in which the IEF, the IEA, and OPEC, together with international statistical agencies, release monthly worldwide oil data. This has undoubtedly contributed to market transparency although it still has some way to go in terms of timeliness and quality. JODI is now being extended to include gas markets. The IEF also coordinates regular workshops at which the IEA and OPEC compare and contrast their respective oil market outlooks. These may not lead to agreement but at least they clarify the differences between them. High oil prices always lead to concerns that financial speculation may have destabilised the market and the troika have also been asked to carry out analysis on interactions between financial and physical energy markets.

2.8. *The Energy Charter Treaty*

The Energy Charter Treaty organisation was founded in the 1990s to promote international energy sector investment in Eastern Europe following the break-up of the Soviet Union. The Energy Charter is intended to enhance confidence in energy investment, transport, and trade, and to reduce political and regulatory risks. It takes the form of a legally-binding treaty with fair treatment provisions. Many nations in Europe, Eurasia, and the Caspian region have ratified it. The Treaty has settled disputes between investors and governments in more than 45 cases.

The Treaty secretariat has been anxious to extend the principles of the Treaty beyond its current members, especially into the Asia Pacific region. In 2015 a non-binding statement of Energy Charter principles, the International Energy Charter, was signed by more than 70 nations and organisations, including most of the major players in world energy, including China and other major Asia Pacific nations.

China has a particular interest in the Charter because of the vast scale of its international energy investments, expected to increase further through China's "Belt and Road" initiative for investment in Central Asia. China already benefits from the fact that some of the countries in which it has made major investments, such as Turkmenistan, have ratified the legally-binding Treaty. Assuming that these countries follow the rules of the Treaty, China's investments in these countries will be better protected. Chinese investors may be able to exercise rights under the treaty provided that they operate through businesses that are registered in countries that are Treaty members. China is actively engaged in the work of the Charter, but has not ratified. China's eventual membership could transform the significance of the Treaty, but it would be a big step for China to expose its relations with international investors to international arbitration.

Russia has a chequered history with the Charter. Russia signed but never ratified the Treaty. At one time Russia committed itself to act in conformity with the Treaty. However, in 2003 the Russian government took back the assets of Yukos, at one time Russia's biggest oil producer, through tax demands that were widely regarded as politically motivated. In 2014 shareholders won a $50 billion award from an arbitration court, sitting in the Hague, under the provisions of the Treaty. The court ruled that Russia's signature of the Treaty implied provisional application at the time of the Yukos action, notwithstanding that Russia terminated this provisional application in 2009. It is not clear whether the court's judgement can be enforced. Nevertheless, it demonstrates that the Treaty is not to be entered into lightly.

2.9. *Regional Bodies*

Many regional and bilateral energy initiatives are also making useful contributions. These include the Energy Working Group of the Asia-Pacific

Economic Cooperation organisation (APEC) and the Association of Southeast Asian Nations (ASEAN) Plan of Action for Energy Co-operation. Such collaborations have made useful, but fairly modest, contributions. For instance, the ASEAN plan of action has made some progress towards the integration of gas and electricity grids in the South East Asia region.[xi] One can also regard the European Union's partially successful efforts towards creating a single energy market as a regional energy initiative.

2.10. *Trade and Finance*

There are many international bodies that have a big influence on energy but are not energy specialists. I am thinking of the World Trade Organisation and various international financial institutions, including the World Bank and its Global Environment Facility (GEF), the IMF, and regional development banks such as the Asia Development Bank. They have important parts to play in the energy scene but none has energy policy as its central focus. They date back to the post-war era, and, as I have already mentioned, they too have had to adapt, with varying degrees of success, to modern realities. There are important parallels with energy, but the evolution of these organisations is beyond my scope.

The rapid increase in trade in clean-energy technologies has become a very live issue at the WTO. The Organisation has been asked to rule, for instance, on anti-dumping measures imposed in the US and Europe on imports of solar panels from China.

3. Need for Change

It is not difficult to point out the weaknesses in this structure of global energy governance. It tends to preserve the divide between developed and developing countries. As a result some of the most important payers, such as China, India, Brazil or South Africa, are excluded from consumer cooperation on energy security and supply. This partly is a question of formal

---

[xi] S. Xunpeng and C. Malik, *An Assessment of ASEAN Energy Cooperation Within the ASEAN Economic Community*, ERIA Discussion Paper Series, December 2013.

mechanisms, but also concerns the mutual understanding and trust that is needed to handle difficult situations.

We are not making the most of opportunities for cooperation between producers and consumers on the efficiency of energy markets. The programme of action agreed at the Jeddah Oil summit of 2008 has been pursued jointly by the Secretariats of the IEA, OPEC, and IEF. Useful work is being undertaken. But this vital process needs stronger direction and support.

There is a lack of engagement between developed and developing countries on the energy policies that will need to underpin new international agreements on climate mitigation, and especially on the crucial topic of energy for economic development. The IEA is not sufficiently oriented to the energy challenges facing developing nations, and this leaves a big gap at the heart of this process.

The limited membership of the IEA, and, to a lesser extent, its history as a counterweight to OPEC, prevent the IEA from supporting the G20 in the way that it has supported the G8 in the past, and this makes it much more difficult for the G20 to provide effective leadership on energy cooperation.

There is a lack of coordination of the many bodies that have been created in recent years for cooperation on specific areas of energy technology and policy. That is why the 2014 G20 Principles on Energy Collaboration aimed for "Enhanced coordination between international energy institutions" and to "minimise duplication where appropriate".

4. The Basis for Cooperation

All governments share broadly similar domestic energy policy objectives. These include security, diversity, and affordability of supply and environmental protection (including climate mitigation). All governments are looking for cost effective ways of improving energy efficiency and introducing advanced and low carbon technologies. There is no reason why cooperation on domestic energy policy could not be extended to include developing countries and perhaps in some areas also OPEC oil producers — bearing in mind the rapid growth of energy demand in OPEC countries. But the world is full of rivalries and political tensions.

Getting the major players around one table, even to discuss topics of mutual benefit, remains fraught with difficulties.

4.1. *Climate Change*

Climate change was discussed at length in Chapter 4. As described in that chapter, the November 2015 Paris summit of UNFCCC created a framework for individual nations to register their planned emissions reductions. This "bottom up" approach depends on the ability of each country to devise and implement low-carbon energy strategies that also meet national aspirations for affordable energy supply and economic development. Much of energy policy is international, so international cooperation needs to play an important part, especially cooperation between rich and poor countries, because it is the poor countries that are generating all the growth in greenhouse gas emissions. The UNFCCC is developing institutions to promote this, including the new Technology Mechanism. But the UN is not the ideal forum for this kind of technical cooperation. The developing nations will need strong international support, including the support of the IEA, to realise their potential.

4.2. *Technology*

Over the years the IEA has built-up a structure of more than 40 international technology collaborations. These are coordinated through the IEA's Committee on Energy Research and Technology (CERT). Technology collaboration has rightly been identified as a crucial dimension of the fight against climate change and, as described above, this has led to the creation of a number of new collaborations in the most important areas. Unfortunately, mainly due to the limited membership of the IEA (and notwithstanding that the IEA's technology networks have themselves been opened to non-members of the IEA), these have generally not been built on the IEA's networks but have, to some extent, duplicated them. There is an obvious need for better coordination of these bodies, and the IEA could host a clearing house, but only if the IEA can establish a wider constituency for this role.

4.3. *Global Markets and International Energy Security*

There is less policy agreement on security and market issues. The IEA is committed to open and competitive markets, whereas OPEC acts as a cartel aiming to manage the market through production quotas. But the relationship is more nuanced than that and need not be simply adversarial. As I have mentioned, at times of extreme price volatility OPEC can be an ally of the West in seeking to calm markets. OPEC is cooperating, through Joint Organisations Data Initiative (JODI), in improving oil and gas market information.

Also, energy demand has been growing rapidly in many OPEC countries. Their domestic energy policies are beginning to converge with those of consumer nations. The practice of providing cheap energy for internal use has led to rapid growth in domestic oil consumption which is beginning to make significant inroads into the volumes available for export. There is a heavy and growing cost in terms of export revenues forgone. So OPEC countries are beginning to take an increasing interest in energy efficiency and, especially in the Middle East, in solar power. In the medium to longer term, as the world reduces its climate emissions, the economies of the OPEC countries will need to become less reliant on oil and gas. Saudi Arabia is already starting to face-up to this. As discussed earlier in Chapter 2, this should be a subject for constructive international cooperation with consuming nations.

4.4. *Emergency Planning*

In June 2017, although prices have steadied following OPEC's agreement with Russia, international oil markets appear well supplied. Probably they will tighten again in the next few years, as relatively low prices continue to reduce investment in supply and increase demand. At that time a market disruption, for instance due to instability in a major oil producing nation, may have the potential to cause an international crisis of energy supply.

Energy consumers will need to be well organised to cooperate to take emergency response measures and to keep international energy markets open. The IEA, founded to deal with the threat of an Arab oil embargo, has the necessary mechanisms in place. All energy consumers stand to benefit from these IEA policies, but their effectiveness is progressively declining along with the OECDs share of international energy trade. China is building its oil stocks, and India intends to do so. There would be obvious benefits to all parties from coordinating the use of these stocks with those of the IEA. This is not just a question of formal mechanisms, but also requires the building of trust and shared analysis of oil market developments between consumer countries over time.

Natural gas was traditionally transmitted through pipelines, making gas security a national, or in some cases regional, issue. But by 2014, 42% of global gas trade was in the form of shipped Liquefied Natural Gas (LNG) and the IEA expect this share to continue to increase.[xii] As a result, gas trade is beginning to assume some of the same geopolitical security risks as oil. There is an obvious need for international planning to cope with a major interruption in LNG supplies, and this is a live topic of energy governance today. It would make perfect sense for this to be lodged at the IEA, provided that the IEA can win the confidence of those major gas importing nations, such as China and India, who are not currently members.

5. An Improved Structure

5.1. *A Grand Bargain?*

In a speech of 13 October 2008, Robert Zoellick, then President of the World Bank,[xiii] called for a "global bargain" between producers and consumers of energy. "World energy markets are a mess". The bargain would include sharing plans for expanding supplies, improving efficiency and lessening demand, assisting with energy for the poor, and considering policies for climate change. "The World Bank can play an important role here". "We might consider taking the global bargain further. There might be a common interest in managing a price range that reconciles interests

---

[xii] International Energy Agency, *World Energy Outlook*, 2016.
[xiii] Zoellick, R.B. *Modernizing, Multilateralism and Markets*, Remarks to the Annual Meeting of the Board of Governors of the World Bank Group, 13 October 2008.

while transitioning towards lower carbon growth strategies, a broader portfolio of supplies, and greater international security".

Zoellick's idea of an international agreement to manage the price of oil is not new and has been proposed countless times in international debate. But I think it goes too far. The first difficulty would be to agree what the price range should be. Everyone says that they want more stable prices but producers and consumers will not necessarily agree what that price should be! More crucially, a price controlling mechanism is unlikely to be effective except, perhaps, over limited periods. It requires some kind of buffer stock which will buy oil when prices are getting too low and sell oil when they are too high. However, the capacity of the buffer is inevitably finite and, sooner or later, the underlying market is likely to get so out of step with the agreed price range that it runs out of money or oil. At that point the mechanism is broken and, as with OPEC's decision not to support the price of oil in 2014, volatility is likely to be even greater than if there had been no attempt at price management in the first place. The governments backing the scheme will probably have lost a great deal of money! One also has to consider the politics, from the point of view of consumer nations, of international efforts to keep the price of oil above its market value and, from the perspective of producers of trying to keep it below market value. Such efforts are not likely to be sustainable for long.

However, Zoellick is right that closer global cooperation is needed on energy data and investment, on energy to advance living standards, and on climate change, and that will require more effective energy governance.

5.2. *An Inclusive Global Organisation*

All this points to the need for a genuinely global body for cooperation on energy policy including all major energy consuming countries and working with energy producers in areas where they have interests in common. It would build trust and mutual understanding between the major participants in energy markets. It would promote predictable legal and regulatory frameworks for international energy investment.

Such a body could support the G20 on energy policy matters and it would provide coordination for the many existing bodies for collaboration on particular technologies. It would coordinate the energy security strategies of all major consuming nations and it would provide a stronger base for the provision of energy market data and analysis. With the inclusion of major developing nations, it would play an important part in articulating low-carbon development options needed to underpin the success of climate negotiations, and it would support the institutions of the UNFCCC.

It would be possible to create such a body from scratch. The membership could be the G7 plus the BRICS (not very different from the old G8 Plus 5, except that all would be equal) or the G20. These ideas have been suggested during some of my discussions in China. But they would be very difficult to implement.

There is tremendous resistance to the creation of major new international bodies. This is partly because of the huge diplomatic effort required to agree their mission, and decide on membership, funding and other aspects. New bodies inevitably overlap with the roles of existing bodies and this is especially so in the case of energy given the existing organisations that I have described. They take a long-time to become fully effective. And governments are reluctant to spend more money on international organisations. National budgets are tight, national foreign services usually consider themselves to be seriously underfunded, and spending on highly paid international experts simpering in gilded halls is a soft target for budget critics.

Sometimes, it is true, new bodies have been founded for the wrong reasons. Governments may press for the creation of new organisations in order to demonstrate their political impact when it would have been better to strengthen existing bodies. That has certainly contributed to the proliferation of institutions in the energy field.

Nevertheless, major new international organisations usually only arise as a result of major crises. The G20 arose from the financial crisis, as I have described, the IEA from the Arab oil embargo, the International Energy Forum (IEF) from extreme oil prices, the Energy Charter Organisation from the break-up of Soviet Russia, and the OECD from the post-war devastation of Europe. In spite of all the challenges that we face, there is no comparable sense of crisis in energy policy today.

By far the best chance of developing fit-for-purpose energy governance today is through strengthening, modernising, and rationalising existing institutions. This means reforming the IEA to make it a more inclusive body and enhancing the roles of other major bodies including the G20, the IEF, and the Energy Charter. It also means better coordination between existing bodies. This faces less political resistance, makes use of existing competencies, and can become effective in a shorter period of time. However, reform of existing bodies can also be a rocky road and if this proves impractical we may yet have to return to the idea of a new body.

5.3. *Reforming the IEA*

There have been many calls for reform of the IEA. Henry Kissinger, who was a leading figure in creating the IEA, said in his address to the IEA Ministerial Conference in 2009, "The IEA now stands at a critical juncture. The world has changed considerably since 1973. In order to be effective in this new landscape the IEA must be prepared to evolve with it".[xiv]

In her 2009 confirmation hearings as US Secretary of State, Hilary Clinton was more specific.[xv] It is worth quoting her statement extensively:

> The IEA should be laying the groundwork now for eventual Chinese and Indian membership in order to achieve the benefits of: (1) Increasing energy policy coordination with rapidly growing energy consumers like India and China; (2) maximising the opportunity for agreeing on energy standards and principles like transparent energy markets; (3) ensuring the coordinated release of strategic petroleum reserves during a major oil market disruption; and (4) maintaining its position as the voice of the world's major energy consuming nations. ...The IEA was created as an institution that represents the interests of major energy consuming nations. If its membership does not change to reflect who those nations are today, its authority and effectiveness will erode.
>
> Full membership would likely require the modification of the original 1974 International Energy Program Treaty Agreement that created the IEA, but the range of options potentially available to integrate China and India into the IEA have not yet been explored. ...The State Department

---

[xiv] Henry Kissinger, *The Future Role of the IEA*, Speech to the 35th Anniversary of the International Energy Agency, Paris, 14 October 2009.

[xv] US Senate Committee on Foreign Relations. Session 1, 13 January 2009. Responses to Additional Questions. Reply to Question No. 133 by Senator Lugar.

will support these efforts, up to and including revision of the International Energy Programme (i.e. the IEA's founding treaty).

Since then, the IEA has developed much closer working relations with major developing countries and the activation of a formal Association with China, India, Indonesia, Morocco, Singapore, and Thailand could be regarded as an advanced stage in exploring closer relations short of actual membership.

It is still too early, in June 2017, to judge the attitude of the Trump administration to the IEA. The President's decision to pull out of the Paris agreement on climate change now makes the role of the IEA, as a forum in which the US engages with other nations on energy policy, all the more important. Rex Tillerson, the new Secretary of State, is a longstanding supporter of the IEA, and that may be an encouraging sign. However, the President's "America first" philosophy and deep suspicion of international organisations generally rather point in the other direction.

What are the implications, and potential barriers to reforming the IEA? To those who are not closely involved in energy governance and, I would say, to the proverbial energy expert visiting from Mars, it seems frankly incredible that the world's leading body for energy cooperation does not include the major developing countries who are such important players. But from within the IEA the difficulties can appear formidable. How would wider membership affect the culture of what is today a rather homogeneous organisation? What voting rights will new members have and will this water-down the rights of existing members? How would the costs of an enlarged IEA be shared? What legal processes would be required to amend the Treaty and does this mean, in effect, negotiating a new treaty, which would be an enormous undertaking?

The voting rights of existing IEA members are still based on oil consumption in 1974, because members have not been able to agree to update them, even though regular updating was expected in the treaty. One should never underestimate the difficulty of getting 29 sovereign nations to agree on anything!

IEA member governments should start with the basic question of what the overall aims of the IEA should be. There are a number of sources to draw on. The first, of course, is the IEA's Treaty itself which is largely concerned with oil security and emergency response. The issues of energy policy are much wider today, but, for all nations, security remains the bedrock. Then there is the last statement of the IEA's aims, the "Shared Goals" formulated by the Governing Board in 1993. These went well beyond the oil security provisions of the treaty of 1974. The emphasis was on free and open energy markets, energy diversity and efficiency, and on environmental sustainability. However, there was no reference, at that stage, to climate change, to renewable energy, or to the energy needs of developing countries. The next source is the two UN Sustainable Development Goals (UNSDGs) that are most relevant. That is to say, those on affordable energy for all, and climate change. And finally, IEA governments could look at the G20 principles. These cover much the same ground as the other sources, with an emphasis on affordable and reliable energy for all. They highlight specifically the importance of high quality data and analysis and of phasing out fossil fuel subsidies.

If the IEA intends to become a truly global energy agency then it should consult widely on the formulation of its new mission statement. It should certainly consult the members of its Association. The mission statement will be the basis of the modernised IEA. Future members will need to be comfortable with it.

The mission statement does not necessarily need to be written into the IEA treaty. Indeed there would be advantage in keeping the statement of objectives in any new treaty as basic as possible to allow for flexibility as new challenges arise. I will come back to the treaty in a moment. But the Agency must be formally committed. The original decision of the OECD Council which established the IEA said that "additional responsibilities" could be given to the Agency. So the mission statement could be implemented in the form of additional responsibilities conferred by the OECD. In any case, it should be a formal decision of the Governing Board of the IEA at Ministerial level.

The IEA treaty itself, the International Energy Programme, was written for a different age and in different circumstances. The signatories believed that they were facing an oil shortage disaster and they wanted to bind themselves legally to a highly specific mechanism for sharing available oil between them. It was a life or death situation and the rules had to be watertight and politically acceptable. As described in Chapter 3 the IEA has maintained oil stocks as required in the treaty, and has released them to good effect on several occasions. However, the specific highly elaborate mechanism for releasing stocks that is in the treaty has never been used. More flexible, and much quicker, methods have been devised.

This elaborate mechanism occupies the bulk of the treaty. The parts of the treaty that continue to be used are those setting up the establishment of the IEA and its committee structure, and the requirement to hold oil stocks, currently set at 90 days of imports. The IEA has implemented stock release and demand reduction quite successfully and very promptly, on a number of occasions. But it has used much simplified mechanisms agreed by its Governing Board. Today the triggering of emergency response is effectively a consensus decision of member countries.

Shortly after the IEA was founded, in 1975, the members realised that it was highly desirable to include Norway, a sympathetic nation with large oil resources. However, as an oil producer, Norway was not willing to bind itself to the emergency oil-sharing rules. The outcome was a formal "Agreement" between Norway and the IEA. This effectively conferred full membership on Norway without Norway having to ratify the IEP or commit itself to participate in the oil-sharing arrangements. The arrangement was not to be a "precedent". But of course it illustrates that almost anything is possible with regard to IEA membership, without treaty amendment, provided that the political will is strong.

Is there a case for rewriting the IEA treaty? A modern international agency, operating mainly by consensus, does not need a treaty anything like as elaborate as the IEP. A new treaty would contain a statement of objectives plus basic rules for the establishment and functioning of the governing board. Everything else would be for the governing board itself to decide, including revisions to the mission statement. Nevertheless, writing a new treaty would be a major diplomatic undertaking. One could argue that focusing diplomatic activity on creating a genuinely global framework for pursuing objectives such as energy security, climate mitigation, and the energy needs of developing countries would not be a bad thing! Probably, however, it is a bridge too far.

The alternative would be to remove just the short phrase in the treaty that restricts IEA membership to the OECD countries. That may not be strictly necessary, if new members, like Norway, are not actually going to sign-up to the treaty, but it is essential to send a powerful signal, not least to prospective new members, that the IEA is serious about reform. The IEA can also change its internal structure, without any change in the treaty, to add top-level committees on climate mitigation and the energy needs of developing countries.

IEA member countries sometimes express alarm about the scale of the voting rights to which very large countries such as China or India would be entitled on joining the IEA. In practice, the IEA very rarely votes on anything and functions as a consensus body. However, the voting system that exists rather favours the smaller countries because most decisions have to be supported not only by a majority of votes weighted according to oil consumption, but also by a majority of members.

There is also concern that opening the doors of the IEA to non-OECD countries will rapidly lead to a huge increase in membership, making the IEA as unwieldy as many UN organisations. There is an inescapable dilemma that the more inclusive the IEA becomes the more difficult it will be to reach agreement. But this is in the IEA's own hands. There is nothing to stop the IEA from managing its membership to give priority to major players on the energy scene.

Oil security is an essential part of the IEAs role. The maintenance of oil stocks and cooperation in emergency response should continue to be membership requirements for countries that are net importers. China is rapidly building strategic oil stocks, though it has probably not reached the level of 90 days of imports that is the IEA benchmark. India is also planning to build oil stocks but is further behind. There may be scope for negotiation on timing and even on whether 90 days is the essential level. Major developing countries are most unlikely to be willing to place their strategic oil stocks under the legally binding control of a body that, for some time at least, is dominated by the West. On the other hand they will find, in practice, that cooperation is in their own interests. There should be room for agreement.

5.4. *The G20*

As has been described, the G20 has geared itself up to be more influential on energy policy with the establishment of its Energy Sustainability Working Group (ESWG), regular meetings of G20 energy ministers, and its agreed principles. It already has worthwhile programmes in a number of areas.

The G20 has called for more representative energy institutions and for better cooperation between them. Through its high level leadership and commissioning power, the G20 can help to bring these things about. For instance, the larger countries who are members of the IEA and most of the developing countries who would be early candidates to join it are all members of the G20. Discussion at the G20, including informal discussions outside the summit meetings themselves, could give real momentum to the opening up and reform of the IEA. This topic is of no interest to Saudi Arabia, but they have no strong reason to be opposed.

The G20 can promote the proposal, referred to above, for an energy efficiency hub at the IEA to be set up in cooperation with the IPEEC. This would be another useful step towards promoting greater coherence in the structure of international energy institutions. More important, it could ensure more effective international cooperation on what is arguably the most important energy policy topic of all.

G20 sponsorship is particularly important for the IEF. The IEF is vital for the relationship between OPEC and the IEA and for oil and gas producer/consumer dialogue. I discuss the contribution that the IEF can make below.

I would like to see the G20 promote a really major global initiative on energy investment for economic development and low-carbon transition. This is needed to promote economic growth and raise living standards around the world and to get us back on track to meet the 2°C climate target. It wouldn't be easy to arrive at an agreement on this because each country, individually, is concerned about its own budget deficit and national debt. A collective effort that successfully raised global growth rates could actually be good for tax revenues and national finances.

5.5. *International Energy Forum and Producer/ Consumer Dialogue*

The IEF is the link between OPEC and the IEA and the forum for consumer/producer dialogue. It coordinates the Joint Organisations Data Initiative (JODI), which, though imperfect, is the best truly global effort to publish accurate oil and gas data and improve the transparency of markets. Unfortunately, today, the IEF secretariat is a bit of an orphan. It is underfunded and not much used by world leaders.

There is a strong case for breathing new life into the IEF. Today, in 2017, there is growing concern that recent cutbacks in investment in new resources, as oil prices have fallen, may be storing up a sharp rebound in prices. A reinvigorated IEF, working with OPEC and the IEA, and with enhanced data from JODI, could shine some light on the supply and demand outlook.

A Chinese national, Dr Sun, was recently elected as the new head of the IEF secretariat in August 2016. Possibly China and India, who will chair the next IEF Ministerial in 2018, will be willing to give a lead.

For the longer term, it is important to remember that the low-carbon energy transition is not only a transition by consuming countries from fossil fuels to clean energy, but is also the transition for producer-nations to reduced reliance on fossil fuel revenues, and the development of new industries. Saudi Arabia is already beginning to face-up to this. There is a need for mutual understanding and cooperation and the IEF could facilitate this.

Today the IEF is under-resourced and little used by world leaders, except as host to their top-level meeting, held every other year. With stronger support and active guidance from the G20 it could make a much more significant contribution.

5.6. *Energy Charter*

To meet the world's needs for affordable energy and achieve the low-carbon transition there needs to be a sound basis for international energy investment. That is what the Energy Charter provides in the regions where it has been adopted. That is mainly Europe, including Eastern Europe, and some parts of Central Asia. I hope that this will spread to Asia and the Pacific Rim and eventually to Africa. Probably, this depends on the leadership of China. As I have mentioned, ratification by China would be a huge step for the Charter. However, China's energy sector is going through a major transformation as state industries are restructured, more competition is brought into the market, and, hopefully, there is increasing scope for international investment. China is not likely to take the risk of exposing its relations with international energy investors to independent arbitration, at least not for many years to come. Nevertheless, China's active participation in the work of the Charter, as a forum for promoting best practice in the regulation of international energy investment is already making a positive contribution.

6. Conclusions and Recommendations

We do not have the most basic tool required to tackle the global challenges of energy policy. That is to say an organisation to which all the main players belong that is focused on the interlocking tasks of security, raising living standards, and climate change.

The G20 has the right membership, but it is a leadership body primarily concerned with economics and finance, and not a working energy institution. The G20 can, however, give guidance to existing energy organisations and lead the reform of energy governance. It has already begun to play this role and it is important that this should continue under the coming presidencies.

Probably we are stuck with the fact, at least for the time being, that OPEC energy producers are in a different camp from the energy consumers and importers. However, the differences are not as great as they once were. OPEC shares with the consumers the aim of improving oil and gas market data and analysis, and of avoiding the most extreme price fluctuations. As energy consumption in OPEC countries increases, so their governments come to share some of the same domestic policy issues as importing countries. Eventually, OPEC countries will have to face up to the consequences of the climate policies of consuming countries, as Saudi Arabia is beginning to do. So there are plenty of areas where OPEC and the consuming countries can work together fruitfully, even if they are unlikely to reach agreement on a number of issues. The best forum for this work is the troika of the IEA, OPEC, and IEF. The G20 should breathe life back into the IEF by funding and commissioning a new programme of troika projects, to include extending and deepening the JODI data system.

A new global body on energy cooperation might seem the most logical answer. But there are serious drawbacks. Most importantly, we don't have the political will, today, for such a major diplomatic initiative. This is partly, in the view of the Western nations, because such a body would overlap with the role of the IEA which is highly effective within its own sphere. But fundamentally, although wrongly, we don't have the perception of a sufficiently acute crisis to galvanise the nations into action. Even if there was political will to create a new body, it would be a long time before it could become effective and the risk is that it would emerge as a politicised body with a very wide membership in which national point-scoring took precedence over practical cooperation.

The alternative is to reform the IEA by increasing its membership to include the major developing nations and by giving it a new mission statement focused squarely on today's energy challenges. The advantage of this is that the IEA already has an effective secretariat and a reputation for working at a practical and technical level, below the radar of political contention. Although still formally an OECD organisation, the IEA has deepened its relations with major developing nations and set itself the task of becoming a genuinely global body.

Adapting the IEA to meet global energy needs will not be an easy process. Member government must be willing for the IEA to modernise and major developing nations, such as China, India, South Africa, Brazil, will have to recognise the IEA as a body that will genuinely serve the interests of developing and well as developed countries alike.

There is a wind of change at the IEA. The Secretariat and the Governing Board understand the need for modernisation and are working to achieve it. The Association with developing countries is a major achievement and, perhaps, a turning point. But there is a long way to go. The success of the IEA in turning itself into the global energy body that the world so badly needs will be critical for meeting our energy challenges.

Chapter 11

Conclusions

1. Introduction

The bedrock of energy policy is secure, reliable, affordable supply. In the West, we already have that, at least for the moderately affluent. It's an extraordinary technical and industrial achievement, but we take it for granted, as we do other utilities such as sanitation and telecommunications. Perhaps we are right to do so. Woe betide the policymaker who loses sight of this fundamental. Even in rich countries there are plenty of people who struggle to pay their electricity bills.

In developing countries, the situation is different. Reliable and modern energy for all is an aspiration and not a fact. One of the problems with the debate on world energy topics is that it tends to be Western dominated, whereas it is the developing regions, such as China, India, parts of Southeast Asia, and Africa that are driving most of the change. Recognising this rather stark fact is the first step towards wisdom. It is in these developing regions that all the growth in energy demand and climate emissions is taking place, whereas demand in the developed countries is static or beginning a gradual decline.

The growth of energy demand in the developing world has been part of a global revolution in living standards. Between 1990 and 2008, the proportion of people living in extreme poverty halved, from 43% to 22%, an astonishing and unprecedented improvement in the human condition. All the major indicators of human development reflect this trend. As the UN has noted, "never in history have the living conditions and prospects of so many people changed so dramatically and so fast".[i] This is an excellent development, which is perhaps not adequately recognised in public debate. It isn't all gloom and doom. But there is still a long way to go to relieve energy deprivation around the world, and this process needs to continue.

The driving force for this change has been the spread of the industrial revolution around the world. And like the original industrial revolution in the UK in the 19th century it has been based to a considerable degree on coal. Indeed, coal is still the cheapest and most accessible fuel in many parts of the developing world. However, coal is also the prime source of greenhouse gas emissions and of local pollution.

As Presidents Obama and Xi Jinping agreed in a memorable 2014 statement, climate change is one of the greatest threats facing humanity.[ii] If we are to limit global warming to 2°C, the target agreed by world leaders, we can afford to emit only a further 800 billion tonnes of carbon dioxide. Today we are emitting about 30 billion tonnes each year and the trend is still upwards. Considering that factories and power stations running on fossil fuels may have lives of 40 or even 60 years, and the inertia of the existing stock of vehicles and buildings, we are at the 11 hour!

Living standards in much of the developing world have risen spectacularly in recent decades. But in the developed world, for ordinary people, they have largely stagnated. They have not shared in the feast. Many ordinary people, in the words of UK Prime Minister May, are only "Just About Managing".[iii] We are living in the age of BREXIT and of president Donald Trump, in which the resentment of the Just About Managing (JAMs) has become a potent political force. We are going to have to find smart policies for speeding up the low-carbon transformation with the absolute minimum impact on the living standards of ordinary people in the developed economies.

So that is the essential conundrum; how to meet growing energy demands around the world while, at the same time, sharply reducing carbon emissions. And how to achieve this at acceptable levels of cost.

---

[i] United Nations, *Human Development Report*, 2013.
[ii] The White House, *US-China Joint Announcement on Climate Change*, 11 November 2014.
[iii] Theresa May, First Statement as Prime Minister from Downing Street, 13 July 2016.

Some will say that this is a totally false dichotomy, because low-carbon renewables, such as solar and wind, together with improvements in efficiency, are not only good for the environment but will also reduce costs and offer the best way to empower developing nations. This may well be the case. After a period of transition, we may indeed emerge into the sunny uplands of energy policy in which low cost renewables have solved many of our problems. For reasons discussed below, I think that caution is appropriate. But what is undeniable is that for this generation, the low-carbon revolution is bound to involve some increased cost and changes in lifestyle. We need to minimise the costs and to ensure, as far as possible, that the changes in lifestyle are welcome.

And there is a spoiler, which is the market price of fossil fuels, especially oil. The volatility of the price of oil, which is around $50 per barrel at the time of writing but which could be anything between $120 and $30 when you read this, is the elephant in the room of energy policy today. This volatility can be highly disruptive and still has the capacity to do considerable damage to the economies of producers and consumers alike. High oil prices fall on ordinary consumers and slows economic development; low oil prices, coming too suddenly, can destabilise producer nations.

These issues are connected. The growth of energy supply in the developing world, and continuing high levels of supply in the West, are not going to achieve the quality of life that is sought if they lead to environmental disaster. On the other hand, if we are concerned about the future impact that climate change will have on the poorest, we should also consider the energy needs of the more than one billion people living in extreme poverty today. The price of oil today may seem remote from the need for a low-carbon transition, but it directly affects the market for renewables as well as fossil energy, and the ability of developing nations to progress.

A former UK Prime Minister Benjamin Disraeli famously described the rich and the poor as, "two nations between whom there is no intercourse and no sympathy; who are as ignorant of each other's habits, thoughts, and feelings, as if they were dwellers in different zones, or inhabitants of different planets".[iv] It sometimes seems that environmentalists and

---

[iv] Novel by Benjamin Disraeli, *Sybil, or The Two Nations*, 1845.

the conventional energy industries are indeed inhabitants of different planets. Often their events and presentations have almost no common ground. The environmentalists seem unaware of the achievements of the energy industry and the vital role that it plays in the world today. Meanwhile, the energy industry seems happy to plan for a future that is totally at odds with environmental imperatives. We need an integrated approach to these problems. Proposals for such an integrated approach have been a central theme of this book.

2. The Climate Treaty

The UN climate treaty, leading up to the Paris agreement of December 2015 was an outstanding human achievement. That is one of the reasons why President Trump's announcement that the US will withdraw is such a disappointment. For the first time, the nations of the world have come together to meet a common external threat. The promises that have been made so far are far from adequate to meet climate targets, but they should significantly improve the climate outlook and there is a process to enable further "ratcheting up".

However, the Paris Agreement has cemented the political reality that sovereign governments are not willing to be constrained by internationally determined quotas for their carbon emissions. So we have a "bottom up" process. Success will depend not so much on climate negotiations, as on the ability of individual nations to achieve their objectives for secure and affordable energy while, at the same time, achieving sustained carbon reductions. This means that the international institutions through which governments engage to consider best practice for national energy policies and technology deployment now have a crucial role.

President Trump has questioned the validity of climate change science and announced that the US will withdraw from the Paris Agreement. His administration has shown itself to be more concerned with the prosperity of the domestic fossil fuel industries than it is with climate change. In June 2017, the full implications are yet to become clear. But there is no doubt that the loss of US leadership, or even positive participation, in this field is a serious setback that will cast a shadow over the Paris process for the next few years.

3. North–South Cooperation for Economic Development

North–South cooperation on energy policy is crucial for the achievement of global energy objectives. The developing nations are becoming the largest emitters of greenhouse gasses. They also have the greatest need for additional energy. For these countries economic and social development is the top–priority. However, it is the developed nations that have many of the financial, technical, and administrative resources. By and large they are already meeting the basic needs of their citizens for affordable reliable energy and, accordingly, climate change is higher on their list of priorities.

In many cases, the quality and stability of a government is the biggest obstacle to progress in energy, as it is for other aspects of development, in the emerging nations. There is a limit to what outsiders can do about that. But international involvement can certainly contribute to the investment framework for energy projects.

In dealing with poor countries, donor nations and international financial institutions should have economic development and poverty eradication as their first priorities. This is consistent with a low-carbon transition because of the huge potential for renewables, hydropower, and geothermal power in many of these countries. Developing nations are generally keen to adopt advanced technologies. But it is unrealistic to expect them to "leapfrog" the West by adopting radical strategies that the developed nations themselves are struggling with.

In some regions, coal is the cheapest and most readily available source of additional power with the added advantage that it uses local resources. In such cases, some new coal power is likely to be needed to meet pressing energy needs. Ruling coal out altogether may undermine the credibility of donors as genuine partners for economic development and lead to the construction of coal power stations that are much less efficient than they need have been.

Solar power can make an important contribution to bringing electricity to rural populations beyond the reach of power grids. Often it is the most economic option. But there are situations where the reliability and flexibility of diesel power is also required. We have to remember that many societies that are largely rural today will probably urbanise rapidly as economic development proceeds. Equal attention needs to be given to the energy needs of growing cities.

In the climate negotiations, the developed nations have promised to mobilise $100 billion p.a. by 2020 in support of the climate policies of developing nations. It is reasonable to ask for greater clarity on how much of this will be grant or concessionary funding, as opposed to fully commercial loans. About a third of this money was due to come from the US. Now that President Trump has announced US withdrawal from the Paris agreement, the questions of how and whether this funding will be made up are bound to assume considerable importance for North–South relations on climate change.

4. The Energy Revolution

Support for renewables has been a major element of government low-carbon policies, especially in Europe. This has been a big success. The huge sums that have been spent to support wind and solar power have brought about spectacular cost reductions, especially for solar. As a result, in the right conditions, these technologies are close to competitive with conventional power, and it is reasonable to assume that they will become fully competitive. Investment has spread around the world and it is now Asia that is the biggest market. 70% of investment in new electricity generating capacity in 2015 was in renewables, an amazing development. I agree with Ben Van Beuren, CEO of Shell, when he said that, in all probability, solar power will eventually become the backbone of our energy systems.

Electricity may well be the energy source of the future. It is clean and convenient at the point of use and could offer low-carbon solutions for some of the uses that seem most difficult to decarbonise, such as transport and heating. There is a lot of interest in the possibility that we may be able to rely almost entirely on variable renewables for our electricity supply.

In Europe, renewables have typically been subsidised by guaranteeing the price of their electricity. When the penetration of renewables was fairly modest it didn't matter very much that they were insulated from market forces in this way. But now that renewables have become big players this practice has become highly disruptive. In the UK and in Germany it has undermined the viability of other energy sources that are essential to maintain supply, forcing government to intervene by the back door to prop them up. As variable renewables increase their share of the market the value of additional capacity declines, because it is only available when there is already an excess of supply. The renewables industry should be actively helping to manage intermittency by adding storage and by linking up with customers able to adapt their demand.

We need large-scale pilots to see if power systems that are very largely based on renewables can be made to work. This is not only a question of technology, it is also a question of the willingness of consumers to adapt, something that is especially hard to predict. These are exciting possibilities, but the outcome is not certain. At the same time, we should invest in the other main options for non-variable low-carbon electricity in regions without hydro or geothermal. That's is to say, nuclear power and fossil energy with the capture and storage of carbon emissions (CCS). The options for smaller nuclear reactors with passive safety features are of great interest. The UK government is supporting this technology. CCS is of special importance because, at the moment, it appears the only available option for decarbonising heavy industry. It also provides an option for retrofitting coal power stations in the future. We would be taking a huge risk to neglect any of these options.

5. Making the Transition

It will be a long time before we have wholly, or even mainly low-carbon energy sources. During the transition period some of the most cost effective options for emissions reduction will come not from switching to fully low-carbon technologies but from improvements within the sphere of fossil fuels. Switching from coal to low-cost shale gas has been one of the main contributors to emissions reduction in the US in the period 2010–2016. The huge programmes that China and India are pursuing to modernise industry and upgrade its efficiency are amongst the most important carbon-reducing programmes on the planet. Indeed energy efficiency, in all its manifestations, is likely to offer by far the greatest potential for carbon reduction in the next few decades.

Efficient gas plant has an important contribution to make, especially where both the heat and power can be used. So has highly efficient coal plant in some circumstances. The question to be asked is "what is the realistic alternative"? We should definitely not invest in coal where the alternative is gas or non-fossil power. But it makes sense to build efficient coal if the true alternative would be an inefficient coal plant. That may sometimes be the reality in some of the most rapidly growing economies. We may have the option of retrofitting with CCS in the future.

6. Getting Real

It is a shocking fact that, since the climate treaty was signed in 1992, world greenhouse gas emissions, far from contracting, have risen by about 50%. Carbon dioxide emissions from energy seem to have stabilised in 2015 and 2016, but there is still no sign of the sharp decline that will be needed. The share of fossil fuels, coal, oil, and gas, in meeting total electricity demand has hardly changed and remains about 80%. How can this be, considering that we hear so much about the progress of renewables?

In part this is due to the sheer inertia of the vast existing stock of fossil fuel plant. Impressive progress is being made through the introduction of renewables for electric power generation. But so far, we are not making comparable progress in the other main energy sectors, that is to say transport, and heating for industry, and buildings. The next stage of low-carbon policy needs to address these sectors. This is likely to require changes to the design of houses and how they are heated, to modes of transport, and to product design and packaging. Some of these changes will bring important collateral benefits. But they will require significant changes to peoples' everyday lives.

Regulation of buildings, industry, and vehicles has an enormously important part to play and transport and has already proved quite successful. For instance, vehicle efficiency standards have contributed to the gradual decline in oil demand in the US and Europe. Today most of the world's construction is taking place in the developing world, especially China. So the enforcement of building regulations in these regions is especially important.

Regulation has to stay within the realms of what is economically and socially acceptable. For instance, radical changes to the design of houses will need to be pursued, initially, through major pilot projects working with builders and marketing experts. As the public become more aware of the attractions of these designs, and of the fuel savings possible, it may be possible to adopt such designs for new homes more generally and to start adapting existing homes. Significant incentives, possibly based around access to building land, will be needed to build momentum.

Ultimately the progressive introduction of carbon pricing will become essential to meet climate targets. As long as the atmosphere continues to be offered as a free dumping ground for greenhouse gasses we cannot make the necessary progress.

Carbon pricing can bring into play the full range of carbon reduction options, and enlist the ingenuity of all the players in the economy to find out which are the most cost effective and acceptable. Industry, including the oil and gas producers, is crying out for a stable and predictable regime of carbon pricing. That will enable them to plan and invest effectively to bring about the low-carbon future. To mangle Jean-Baptiste Colbert's famous metaphor about taxation, the art of carbon reduction consists in so plucking the goose as to obtain the largest possible amount of feathers with the smallest possible amount of hissing. Carbon pricing can help to achieve that.

Carbon pricing can be introduced in different ways. It makes sense to start with a relatively low-price and announce a firm intention to increase this progressively. A large carbon tax introduced too quickly would hand a big subsidy to low-carbon producers, most of whom have already been subsidised. The aim should be to incentivise future investment and behavioural change and for that it is the future level of any tax that is important.

Initially, a carbon price does not necessarily require a tax at all. Governments can simply commit to pursue regulatory and other policies across the board that assign a specific value to carbon emissions reductions. Carbon trading schemes also have the potential to set carbon prices and they have the advantage of incorporating actual targets for carbon reduction. But the price levels in the European trading mechanism, the most substantial to be introduced so far, have proved rather erratic. Carbon taxes themselves need to be part of a fiscal restructuring that ensures that there is no overall increase in the burden of taxation and affords compensation to those most affected, especially the poor and energy-intensive industry. Without a clear link between the carbon tax and these compensating measures the impact on living costs and on industry may prove intolerable.

It is vital to persuade all concerned that the government, and indeed future governments, will indeed stick to the long-term plan. There is no sure fire way of achieving this. One way to give greater assurance would be to establish extra-governmental watchdogs, on the lines of the UK's Committee on Climate Change. Another is to enshrine the concept in international agreements.

Carbon pricing is politically difficult. That is because it means facing up to the far-reaching and across the board challenge of reducing carbon emissions. Energy will be more expensive. Individuals and businesses will face strong incentives to modify behaviour. That is the reality of meeting the challenge of climate change.

7. Cities and Transport

Most energy is used in cities and rapid urbanisation is one of the most dynamic aspects of the world economy today. Clean and efficient cities can make a big positive contribution to solving our energy problems. This is partly about the efficient design of new buildings. But transport is just as important. Air pollution, partly caused by traffic, has become a major problem. In cities where most people travel by public transport, there is public support for restrictions on polluting vehicles. City governments should give a lead by putting increasingly heavy restrictions on, and eventually banning, oil-fuelled cars from city centres. Cities will become much nicer places, and the take-up of electric, and possible hydrogen, vehicles will be given a boost. This is one of the areas in which, I believe, public opinion will support aggressive climate policies.

8. The Oil and Gas Industries

So far, carbon policies have had relatively little impact on the international oil and gas industries. The big shake ups have arisen from the spectacular growth of shale oil and gas production in North America and from the growth of demand in the Asia Pacific region. The cycle of price volatility continues. Falling oil prices lead to sharp cuts in investment which, in turn, lead to the next round of price spikes.

As long as we are still dependent on fossil fuels, the volatility of their prices retains the potential to disrupt economic progress, and especially the progress of developing nations, for whom oil imports are a big drain on hard currency resources. We must continue our efforts to make the market as open and transparent as possible and to improve the quality of data on future demand and supply. It is important for international institutions such as the International Energy Forum (IEF), the International Energy Agency (IEA), and Organization of the Petroleum Exporting Countries (OPEC) to cooperate on this.

The IEA runs the main international mechanism for dealing with oil supply shortages. All its members are required to hold strategic stocks and the IEA coordinates their use in case of emergency. Unfortunately, the IEA's membership is too limited to manage energy crises effectively in the modern world.

Environmentalists are calling for investors to divest their holdings in oil and gas companies. That is pointless. It won't have any impact on carbon emissions. Oil production will fall as we improve the efficiency of conventional vehicles and, progressively switch to electricity or possibly hydrogen, and not before.

It is reasonable for investors to keep an eye on how their companies will cope when this starts to happen. However, the big concern here is not so much the "major" international oil companies. They have shown themselves adept at handling much more rapid change, and can ultimately return their assets to shareholders. Rather, it is the nations that have become most dependent on oil exports and their national oil companies, which now dominate world oil production. We may not welcome the huge rents that they are able to extract from oil markets today. However, if the loss of these revenues is too sudden, it will undoubtedly exacerbate the instability that we are already seeing in the Middle East and elsewhere. The low-carbon economy of the future will be a difficult transition for producers as well as consumers.

9. Research

Research, research, research. It is technically possible to achieve our energy goals with existing technologies, but it is probably not politically possible. We need further technical progress to increase the accessibility of modern energy and to reduce its cost and carbon content. Critical areas include, PV, electricity storage, and CCS. We should also be working on smaller nuclear reactors with passive safety features, and demand side management systems. Research is probably more cost effective than deployment subsidies and we can afford to spend more.

Basic research and small-scale demonstration is best pursued through individual government and commercial laboratories with flexible international dialogue. But there could be big gains from delivering major demonstration projects for technologies such as nuclear power and CCS through internationally financed and managed projects.

10. Energy Governance

International cooperation on energy policy is vital to achieve our energy objectives. Especially important is the cooperation between developed and developing nations on meeting development needs in a sustainable way. Cooperation is also needed to maintain energy security and coordinate energy research. The IEA remains, by far, the organisation best fitted to these roles. The success of the IEA in meeting its ambition to become a genuinely global body is, therefore, vital for our energy futures.

11. And Finally

I am optimistic about the future of world energy. The continuing spread of modern energy around the world is a success story. But we certainly face difficult problems. We are not on course to solve them today. But there is now a deep-rooted desire for change and there are powerful trends that can work in our favour. These include the advance of technology, new business models, urbanisation, and the steady evolution of modern economies towards a service-based and less energy-intensive model. Once we reach a tipping point, change may be more rapid than we expect.

The energy policy of President Trump is a reaction from these trends. He represents a significant body of opinion in the US, and perhaps elsewhere, which is not to be dismissed lightly. The low-carbon revolution imposes costs on energy consumers and involves disruption in energy markets that is threatening to some. If we are to retain political support for the profound changes that are needed in the energy sector these aspects have to be managed carefully to protect the most vulnerable.

Leading nations around the world, as well as many states and major businesses in the US have vigorously repudiated President Trump's announcement of withdrawal from Paris agreement. We must hope that this pressure of opinion, combined with the realities of the situation, will eventually bring the US back into the fold.

In this book I have called for an integrated approach to our energy conundrum, recognising that the question of climate change cannot be separated from the question of how we will meet the energy needs of the developing world, or from the basic need for reliable affordable energy supply. I have urged pragmatic and realistic policies that avoid too narrow a model of the future and minimise the burden on living costs. I have shown how we can make the absolute maximum use of policies that go with the flow of industrial and economic efficiency, that opens up lifestyle opportunities that people find attractive, and that uses markets and clear pricing signals to create the widest possible business opportunities. I have urged the vital importance of researching advanced technology. I have also highlighted the need for closer international cooperation on energy policy and the institutional reforms that can make this possible.

The prize is immense. We are aiming for a clean energy revolution, an energy access revolution, and a smooth transition. It's not impossible. With the right policies, and a successful drive on technology, it can be within our grasp.

Annexure 1

Case Study: Energy Minister of Ygrene

You have recently been appointed the Energy Minister of Ygrene, an (imaginary) developing country with a population of 100 million and average income of $3,000 per head. Life expectancy at birth is 55 years, infant mortality is 50 per 1,000 live births, and 20% of the population live in extreme poverty.

Ygrene has plentiful coal reserves. Coal supplies 60% of Ygrene's energy and 90% of its electricity. About half the population of Ygrene, living mainly in the North, are subsistence farmers on very low incomes. Many of them cook on indoor wood or animal dung fires and breathing smoky air is one of the greatest health problems. Their lives could be greatly improved by the provision of clean cooking facilities. They also need piped fresh water and irrigation for their lands. But the possibilities are constrained by lack of electric power. Only 71% of the population have access to electricity.

Rapid economic growth in recent years has led to the development of a significant urban middle class, mainly located in rapidly growing cities in the South of the country. These are by far the most vocal and politically influential part of the population. Car ownership is rising rapidly. Both electricity and oil are heavily subsidised by the Government, ostensibly as a social measure.

You are very excited by this appointment. You have long recognised that the Ygrene energy sector is ripe for reform and you are aiming to make big economic, social, health, and environmental improvements. It is your first Cabinet post, but you are still young enough to aspire to even greater things. The Prime Minister is due to retire in a couple of years and, if you can build sufficient political support within the Party, you could become a serious contender to succeed her. You also hope to gain a strong international reputation in the area of climate mitigation.

You have just returned from your first cabinet meeting. The Prime Minister and the Environment Minister arrived hot foot from an international climate change meeting where they have gained international praise for committing Ygrene to a 20% reduction in carbon intensity by 2020. Everyone around the Cabinet table agreed that this must now be a top national priority.

The Prime Minister also raised the question of energy security, about which she is very concerned. Ygrene has no oil of its own and imports are rising rapidly. The government has just completed the construction of the Ygrene Strategic Oil Storage Facility, and the Prime Minister has announced that by 2020 Ygrene will have 20 days of imports in store. Everyone agrees that this is vital. However, filling the Facility is a matter for the Energy Ministry and the cost will come from your budget.

There was a discussion of the Ygrene Development Plan. A key element is the Prime Minister's Warranty to the people of Ygrene, given at the last election, that 95% of the population will have clean cooking facilities and access to electricity by 2020. Everyone agrees that this commitment must be kept if the Party is to have any credibility at the next election.

At the Cabinet meeting the Finance Minister took the opportunity to stress that the country is in a very difficult situation financially and that there will be absolutely no flexibility in your budget. Furthermore, energy prices are uncertain, and should there be any increase in international oil prices you will have to absorb the costs.

The Home Affairs Minister raised the question of the Ordyh Hydroelectric Dam project. The Government has taken no decision on this, and this will be primarily a decision for you. He asked for agreement that should the dam go ahead the arrangements for resettling the local population must be a matter for his Department with agreed funding transferred from the Energy Ministry. Of course, the arrangements will reflect the many other pressures facing his Department. This is agreed. But you are a little concerned afterwards that perhaps you have missed a trick here.

The Trade Minister has reported progress on his negotiations with neighbouring country Edart on a free trade area between your two countries. 25% of Ygrene's exports go to Edart, so this is important for Ygrene's future prosperity. The discussions are at a critical stage, but the Trade Minister is optimistic.

Finally the Foreign Minister was pleased to report a considerable increase in promised development aid from Europe and the US following his visits to Brussels and Washington. As usual, however, there are a range of international policies on which Europe and the US are seeking Ygrene support, especially in the battle against human rights abuses.

As you leave the Cabinet meeting the Prime Minister takes you aside into her private office. She is delighted to have you in the Cabinet and she is sure that you will be an excellent colleague and a safe pair of hands. She reminds you that the reason why the post has opened up for you is the abrupt dismissal of your predecessor. His announcement of the cancellation of all energy and fuel subsidies had led to rioting in urban centres and a collapse of support for the Government. It was either the Government or him. The Opposition have given an absolute assurance of no change in the subsidies for the next 15 years.

The Prime Minister also tells you that leading figures in the energy industry (Electricity and Oil and Heavy Industry) had initially expressed some concern about your appointment. She has privately assured them that you are a good friend to industry and they have nothing to worry about. She is sure you realise how influential they are in the Party. Oh, and good luck!

When you arrive at the office there is a telephone call from your mother. She is thrilled about your appointment. Now at last there is a Minister who understands the problems of farmers and she is looking forward to the extension of power to the North of the country, including the family farm where she lives. She is worried about the health of her grandchildren, some of whom live in smoke filled homes using traditional cooking methods.

You next summon the senior officials of the Department to instruct them on your reform plans. But before you can start the Permanent Secretary, who is a non-party official with many years of experience in the Department, suggests that perhaps it would be better if he was first to brief you on the current situation.

He begins with a general observation. He and his colleagues are enthusiastic about your appointment because they think you are on the side of reform, which is badly needed. Everyone knows what needs to be done. But there are severe constraints. These are partly financial, but they are also political. The budget is tight, and could be even tighter if oil prices go up. The Energy Minister, especially a brand new one, is not the most powerful figure in the Cabinet, and there are powerful political forces in the country. You can afford to take on some of these forces to a certain degree, but you should not take on the most powerful forces head on (as your predecessor tried to do) and you cannot take on everyone at once. Otherwise you are toast.

The budget for 2018 (set by the Finance Department) is $Y700 million. Of this amount, on current policies, $Y400 million is taken up by oil ($Y200 million) and electricity ($Y200 million) subsidies. The main beneficiaries are, of course, the middle classes who use most of the electricity and oil. You can probably get away with making some reduction but anything too sudden will bring people back into the streets. The subsidies do benefit the poor to some extent and if you want to protect them from the effects of any cuts you will need to reinvest a quarter of your savings in social programmes.

Officials have calculated that in order to achieve the carbon-intensity target announced by the Prime Minister, Ygrene will need to make reductions of 60 million tonnes of CO_2 p.a. by 2020 as compared to business as usual. On a flat rate basis that requires a saving of 15 million tonnes in 2018. All this is in the public domain and the Department has also published its calculator of how the savings attributable to each year's programmes are to be measured.

So you should aim to save 15 million tonnes p.a. CO_2 on this basis in 2018 if you want to be deemed to be on course. Of course, you could save less in 2018 and announce an acceleration of savings for the future. But how credible will that be, especially as the average Energy Minister only lasts for two years?

By far, the cheapest way to make CO_2 savings on the supply side is the Ordyh dam. This will save 7 million tonnes of CO_2 p.a. which you can start claiming immediately the dam is announced. And it will cost you nothing. The economics of the dam are so attractive that international investors will pay for it entirely, including generous funds to resettle the local population of about 2,000 people. However, there are problems. The dam requires flooding a mountain valley of great natural beauty which also has a significant population. Although the local people would benefit from a high quality resettlement programme they are deeply suspicious of the government. Protests from them are likely to trigger widespread international criticism. It is vital that you keep firm control of the arrangements so that you can reassure them. The other problem is that the Revir river on which the dam is located flows down to Edart. The Edart government is deeply anxious about the implications of Ygrene control of the headwaters and has threatened to call off trade negotiations if the dam goes ahead. A possible tactic would be to offer Edart a share of the cheap electricity, but, for instance, assigning one quarter of the power to Edart would cost you $Y15 million p.a.

Your other supply-side options for meeting the carbon intensity target are renewables (wind, solar, geothermal) at $Y10 per tonne of CO_2 saved or nuclear power, also at $Y10 per tonne of CO_2 saved. There is no great opposition to nuclear power in Ygrene at the moment, but you are a bit concerned about the weakness of Ygrene's safety and regulatory authorities.

You can also save carbon by increasing the efficiency of Ygrene's fossil plant. 50% of Ygrene's carbon emissions arise from the power sector and (although some new plant is of very high quality) there is huge potential for efficiency gains through closing of older, less efficient plant. A big programme of new, Ultra Super-Critical coal plant, most of which would be paid for by the private sector, would save emissions at a cost of only $Y5 per tonne of CO_2 to the Department and underpin Ygrene's power supply for the future. It is your cheapest carbon saving option after the Ordyh dam. The Industry Minister is very keen on this. But the Environment Minister has told you that he does not think this is the right answer because it will leave Ygrene with a new fleet of coal power stations (albeit efficient ones) with operating lives of 60 years. This may save carbon in the short-term, but it will be a barrier to the low-carbon future that he is planning for Ygrene in the longer term.

For some years, officials have been studying the options for a carbon tax or a carbon cap-and-trade system covering the power and heavy industry sectors. They have concluded that a carbon tax makes the most sense, being the simplest option. This would be a revenue earner as well as saving carbon. A carbon tax would give you a defined revenue stream of $Y 25–50 million p.a. depending on the level you set and save (pro rata to the level set) 2.5–5 million tonnes of CO_2. However, the possibility that you might introduce such schemes is the precise reason why industry leaders were so worried about your appointment. A carbon tax at anything like $Y 50 million p.a. would be taken as a direct challenge by the industry. But you could consider imposing the tax at a lower rate or you could exempt the most sensitive (i.e. politically powerful) industries. This would please the Industry Minister, especially if you exempted the industries that are most subject to international competition, because he is very worried about the impact of carbon policies on Ygrene's competitiveness and growth. Of course your receipts will be reduced by whatever share of the emitting industries you decide to exempt.

Achieving the Prime Minister's Warranty means bringing electricity to an additional 3 million homes per year. In 2018, you can extend the grid relatively cheaply to another 1 million homes at a cost of $Y50 per home. But extending the grid further, to more distant and less densely populated areas is more expensive, an average of $Y100 per home for the remaining 2 million homes needed to keep the programme on track. Alternatively, you can bring electricity to these 2 million homes with "island" projects, mainly small solar, micro hydro, and bio, that will provide power for at least one light bulb and a TV. That will only cost $Y50 per home and it will meet the letter of what the Prime Minister has promised. This will also reduce carbon emissions at a rate of 1 tonne CO_2 per $Y25 spent. However, the Industry and Home Affairs Ministers have urged you to extend the grid as far as possible because that will create the potential for improved agriculture and light industry in new areas.

The warranty also requires that you bring clean air cooking facilities to 3 million additional homes each year. This means installing biogas,

LPG or advanced biomass stoves with good ventilation at a cost of $Y50 per home.

Filling the Ygrene Strategic Oil Storage Facility at 1 million barrels p.a. will cost $Y100 million in 2018, at expected oil prices. This is something that you could defer to later years, but the Prime Minister will not be impressed and the Department runs the risk of higher costs if oil prices rise in the future.

There is one piece of possible good news. The Ygrene National Oil Company (YNOC) has been in secret discussions with the government of one of the major oil producing countries, Ograbme, who have offered a secure supply of oil, sufficient to meet all of Ygrene's needs, at only $Y75 per barrel, 25% below market levels. Accepting this would enable you to cut the oil subsidy by 50% and take 25% off the cost of filling the Storage Facility. The reason for this offer is that Ograbme is having difficulty marketing its oil as a result of an oil embargo imposed by most of the major oil importing nations as a sanction against severe human rights abuses. These discussions have hitherto been kept secret from your Department. You and your officials have read about what is going on in Ograbme and at first you are shocked that such contacts have occurred. But can you resist a proposition that might solve a lot of your problems?

You spend a sleepless night pondering the problems that you face. At 3:00 a.m. you call the Permanent Secretary (he is used to such calls). By exactly how much does he think you could cut the subsidies without suffering the same fate as your predecessor? There is a pause. "I don't think you would get away with cutting the oil or the electricity subsidy by much more than 10% in any one year".

You realise that your new job is going to be much tougher than you thought. You will have to think very hard about your priorities and make some compromises.

Consider how you are going to tackle the challenges that you face as Ygrene Energy Minister and:

1. *Draw up the Energy Ministry's Budget for 2018 and*
2. *Prepare the speech that you will give to the Ygrene Parliament (and the world) announcing your five year energy strategy.*

2018 Budget Analysis for Ygrene Case Study

	Budget $ 700 m	CO_2 reduction target 15 m tonnes	Oil stocks target 1 m barrels	Access to electricity target 3 m homes	Clean cooking target 3 m homes
Cost of oil and gas subsidies $400 m less any reduction					
Save 50% if you accept Ograbme oil					
Cost of using 25% of reduction to compensate the poor					
Ordyh dam cost zero CO_2 saving 7 m tonnes CO_2					
Assign some dam electricity to Edart, one quarter share costs $15 m					
Renewables $10 per tonne CO_2 saved					
Nuclear $10 per tonne CO_2 saved					
Efficient coal $5 per tonne CO_2 saved					
Carbon tax 1 m tonnes CO_2 saved per $10 m in tax					
Extend grid to first 1 m homes $50 per home					
Extend grid beyond 1 m homes $100 per home					
Island electricity extension $50 per home plus 1 tonne of CO_2 saved per $25 spent					
Extend clean cooking $50 per home					
Fill strategic oil reserve $100 per barrel, or $75 with Ograbme oil					
Total cost and achievements towards targets					

Annexure 2

Case Study: Energy Minister of Rewop

Introduction

You are the Energy Minister of Rewop, Rewop is a rich, developed country with a population of 65 million with an average income of US$40,000 per person. Rewop is in a temperate zone, and although there is considerable and growing demand for air cooling in the summer, peak electricity demand, for the time being, is for lighting and heating in the winter. Rewop has no opportunities for hydroelectric or geothermal power.

You are already pursuing sustainable energy policies. You have put in place tough regulations to improve the energy efficiency of homes, industry and vehicles. You have been supporting renewables on a large scale. Coal used to be the main source of electric power in Rewop but you are phasing it out. These policies, which are generally popular, are having a big effect. Today, in 2018, Rewop is emitting 500 million tonnes p.a. of carbon dioxide from energy. By 2050, this is expected to have declined to 260 million tonnes, even though Rewop's GDP is expected to have grown by 50% by then.

A Meeting with the Prime Minister

You are unexpectedly called to the Prime Minister's office. She has agreed to chair the next world climate summit (Conference of the Parties to the United Nations Convention on Climate Change) which will be held in Rewop. Governments around the world have made important commitments to reduce greenhouse gas emissions. But so far these are not nearly sufficient to achieve the target, set by world leaders, of limiting global warming to two degrees Celsius. The Prime Minister believes that it is time to get a grip, under her leadership. Rewop must show the way.

Experts have estimated that, to play its part in a global 2°C plan, Rewop must reduce its carbon emissions by 80% by from 1990 levels by 2050. This means that it must reduce them to 120 million tonnes in 2050. The Prime Minister isn't absolutely sure if this is achievable at acceptable cost, but she needs your advice urgently because, if possible, she would like to announce this target in a pre-summit speech that she is delivering next week.

An election is due in a year's time and the Prime Minister says that it is essential that your plan should not undermine, and if possible should enhance, her chances of re-election. If the opposition win it will be a disaster for the environment because they do not believe established climate change science.

Although Rewop is prosperous by international standards, the economy has experienced a hollowing out of well-paid blue collar employment. For many in the lower and middle income ranges real incomes have stagnated for several decades and life is difficult. Household energy bills have been rising recently, increasing the cost of living, and this has become politically sensitive. Industry blames the loss of employment on fierce international competition and is lobbying vigorously against environmental levies. In these circumstances, the Prime Minister would prefer to avoid anything that looks like a new tax, if that is possible. She also asks you, if at all possible, to limit the total new burdens on energy costs in 2019 (including any carbon tax) to P$10 billion or less. That's about P$150 for every person. She also needs your judgement on how adaptable Rewop's citizens will be to the changes needed for a lower carbon economy.

You have very little time to prepare your plan. It will have to be a "back of the envelope" effort, concentrating on the really big numbers.

You realise that there are many other significant factors, (for instance aviation and shipping) but these will have to wait for later refinement.

There are huge uncertainties about how technology and patterns of energy consumption will develop to 2050. It is unrealistic to make detailed plans, year-by-year until then. Instead, you must make specific spending plans just for 2019 that you believe will set Rewop on a credible pathway towards meeting the lowest realistic emissions target for 2050. Although you will receive expert advice from your team, in a number of critical areas you are going to have to make your own judgements.

The Business as Usual Outlook

The split of the 260 million tonnes of carbon dioxide emissions expected in 2050 under current plans is as follows:

	Million tonnes
Electricity generation	50
Transport	100
Business (industry, etc.)	50
Buildings (heating homes, etc.)	60
Total	260

Also under current plans, Rewop will generate 350 units of electricity (terawatt hours) in 2050 made up as follows:

	Units of power	% of power	CO_2 emissions (million tonnes)
Coal	0	0	0
Gas	100	29%	50
Nuclear	100	29%	0
Renewables	150	43%	0
Total	350	100%	50

A Carbon Tax?

You need help from your advisers in the energy department, and you start to summon them in turn. Your first visitor is your senior economist, Martin. He is shocked by the ever increasing government intervention in energy markets. How can the government possibly predict future technology and cost developments? If Ministers (i.e. you!) make the decisions they will just follow political fashion. The obvious way to reduce carbon emissions is to put a price on them through taxation. In that way the market will, over time, find out the most cost effective solutions, drawing on the widest range of possibilities. These include improved energy efficiency, probably the biggest carbon saver of them all, as well as new supply technology. Carbon reduction will be achieved at the least cost to energy consumers. Martin recognises that the government has an important role in R&D and in the earliest stages of commercialisation of new technologies. But otherwise, government intervention is a recipe for waste and inefficiency.

You can certainly see the logic of Martin's intervention. You do want to reduce carbon emissions in the most cost-effective way. But taxes are unpopular. Governments that introduced carbon taxes have fallen. The Prime Minister has specifically warned against them. Even if you could persuade the general public, the industry lobby will be incandescent! It is going to be a lot easier for you to sell exciting new renewable technologies than a new tax! Has he lost his political head?

Martin says that he realises that this is difficult, but sometimes politicians have to be brave. Are you seriously interested to reduce carbon emissions or not? The tax can be introduced at a very low level and gradually increased, because the aim is to influence investments and decision-making for the future. As low-carbon options are implemented, fewer people will actually pay the tax, so the burden may not be so great. The proceeds of the tax can be used to compensate those who have to pay it, and especially the poor, since a carbon tax would be regressive.

Do you think that you can persuade the citizens of Rewop to accept a carbon tax? Can you convince investors that the government will stick to this commitment through to 2050? If so, you have the option of imposing a tax on electricity generation and industry that starts from P\$20 per tonne of CO_2 in

2019 and rises to P$100 per tonne of CO_2 by 2050. The tax will raise consumer costs by P$400 million in 2019 (but of course much more in later years).

If you impose this tax (and if investors are convinced that the policy will stick), it will make a big difference to emissions from industry and electric generation in 2050.

Total emissions will fall to 210 million tonnes in 2050, made up as follows:

	Million tonnes
Electricity	25
Transport	100
Business (industry)	25
Buildings (homes)	60
Total	210

Electricity generation will fall, to 300 units, as electricity is used more efficiently, and gas will be squeezed, leading to this outcome:

	Units of power	% of power	CO_2 emissions (million tonnes)
Coal	nil	nil	nil
Gas	50	17%	25
Nuclear	100	33%	nil
Renewables	150	50%	nil
Total	300	100%	25

Transport — Electric Cars

Transport is the biggest source of emissions, and a real headache. The largest element is road vehicles. The projected 200 million tonnes of emissions in 2050 would have been much larger but for the tough vehicle efficiency standards that you already have in place, as well as major existing programmes to enhance public transport and get people to walk and bicycle more.

Your transport adviser, Joe, comes bounding into your office to tell you that electric cars are the coming thing. He owns one himself and is delighted with the performance. He expect a steady improvement in the cost-effectiveness of electric batteries. He points out that to make significant further inroads into vehicle emissions you are going to have to persuade the citizens of Rewop to switch to electric cars.

You have one or two questions for Joe. Surely today's electric vehicles are considerably more expensive than petrol vehicles, and still subject to some degree of "range anxiety". They only account for 1% of the market. You realise that this is growing, but not as fast as the manufacturers had hoped.

Joe is confident that consumer preference is about to turn to electric cars in a big way. All the most influential "opinion formers" that he knows are planning to buy one.

You haven't seen very much sign of that yet. However, there is one ray of hope. The capital city of Rewop has a terrible problem of air pollution. The mayor is considering restricting the use of petrol vehicles within city limits and banning them outright from 2030. He will need your support for such a tough move. It would certainly kickstart the market for electric cars. But there will be squeals from low income people who rely on cheap old cars to get to work.

You have the option to spend P$2 billion p.a. on a major programme of subsidies of electric vehicles plus the installation of fast chargers across the road network. This will lead to 30% of cars being electrified by 2050, reducing emissions by 30 million tonnes and increasing the demand for electricity by 25 units. If you are optimistic about the potential for electrifying vehicles, and if you are willing to back the Mayor's tough policy, you can spend P$5 billion to achieve 60% electrification by 2050, saving 60 million tonnes of emissions and increasing electricity demand by 50 units.

Buildings — Electrification of Housing

Buildings are the next source of emissions to consider. Almost all the citizens of Rewop heat their homes with gas or oil. You already have programmes in place that will gather the "low hanging fruit" for improving the efficiency of homes; that is to say, roof insulation, draught exclusion, double glazing, insulation of cavity walls.

The next visitor is Susan, your housing adviser. She says that to make further progress you are going to have to persuade people to switch to low-carbon electric heating. Heat pumps can be used to gear-up the efficiency of electric heating at least three-fold, by drawing on the heat that is in the ground, like a reverse refrigerator. Provided that homes are heavily insulated to minimise the needs, heat pumps can be quite effective. They may be expensive to install, but eventually these costs can be recovered because so much less energy is needed.

You are concerned about the difficulties of persuading existing home owners to undertake this conversion on a really large scale. But new houses may be easier. About 1% of Rewop's housing stock is replaced each year, so about a third of the 2050 housing stock is yet to be built and it seems realistic to concentrate first on these new houses. Even so, there is bound to be some increase in the cost of new houses, which is politically sensitive, and there may also be cultural resistance to the style of super energy efficient homes. Rewop citizens are rather traditional in their attitudes to housing.

How optimistic are you that you can persuade existing or future Rewop home owners to switch to low-carbon electric heating? You have the option to spend P\$2 billion in 2019 to ensure that all new homes have electric heating. This will lead to about 30% of homes being electrified by 2050, saving 20 million tonnes of emissions and increasing electricity demand by 20 units. If you are more optimistic that a significant share of existing homes can be electrified you can spend P\$4 billion in 2019 to reduce emissions by 40 million tonnes with an increase in electricity demand of 40 units.

Carbon Capture and Storage (CCS)

Now you are having a session with a fossil fuel expert, Bill. He believes that the best solution to the climate problem is to continue with coal and gas power generation, but to take the CO_2 emissions and pump them back underground, where the carbon originally came from! The technology is known as Carbon Capture and Storage (CCS). Bill points out that CO_2 is already pumped into underground oil reservoirs on a large scale in the US, to improve oil recovery. There is good evidence that the CO_2 will stay underground.

You are really worried about CCS. You realise that all the components are proven. But today there are very few full scale fossil power stations whose carbon emissions are captured and stored. Some of these, like nuclear, have also run into serious construction problems. CCS reduces the efficiency of a power station by up to 20% and the public has to be persuaded of the safety of pumping and storing CO_2 under high pressure. CCS is not a popular choice with the environmental lobby.

Bill points out that for some heavy industries that require very intense heat, there is, today, no alternative to fossil fuels, so CCS is the only available option for getting their emissions down to really low levels. Also, on a global scale, for some major developing countries, coal is the cheapest and most accessible form of energy so CCS may well be critical for containing world emissions. Rewop could make an important contribution towards commercialising this technology. It is true that CCS is expensive today, but it is still cheaper than renewables plus storage and costs will no doubt come down with development.

You have the option to spend P$2 billion p.a. in 2019 to launch a programme of CCS. This will deliver 25 units of power in 2050 and will also reduce business emissions by 5 million tonnes. If you are more optimistic about what CCS might achieve, and think that costs could be brought down in a major programme, you can spend P$4 billion to produce 50 units of power and reduce business emissions by 10 million tonnes. If you have a carbon tax in place CCS will appear more competitive and the government contributions needed will halve, to P$1 billion and P$2 billion respectively.

Renewable Electricity and the Flexible Grid

The next adviser to come into your office is the renewable energy expert, Ben. He is full of excitement. Technological progress will solve your problem. You have a wonderful opportunity to decarbonise the nation by going over to renewables, mainly wind and photovoltaics. The cost of renewables has been coming down dramatically in recent years and this is expected to continue. Today, in the best locations, they are no more expensive than conventional power and they could well become cheaper.

You ask Ben about the problem of intermittency. How, can you be sure that power from renewables will be available when needed? Ben says that there is a revolution in store. Smart power systems are coming down the track. There are several elements. New international grid connections will enable us to access power from more diverse sources. Sophisticated meters will enable people to adapt their use of electricity to match available supply. Businesses will make the best use of their back-up generation and operational flexibility. Consumers will have lots of options to run appliances, heat their homes, and, eventually, charge their electric cars in responsive ways that do not affect the quality of service. Electricity storage has in the past been prohibitively expensive, but the costs has come down dramatically from $3,000 per unit in 1990 to $200 today. It is still much too expensive for any kind of baseload and there is almost no storage on the system today. But it already has a growing role to meet short term supply gaps.

Ben is expecting a revolution in the way electricity is sold, with new "service" companies using advanced IT to optimise all these demand and supply options.

He also reminds you that not all renewables are intermittent. About 10% of Rewop's renewable electricity comes from biogeneration, which is just as reliable as conventional power. However, this share cannot be increased because of limits on the plant biomaterial that is available.

You press Ben a little further. You have a lot of sympathy with this approach, but, you need to be sure that the lights won't go out. Today, with only 15% of electricity supplied from renewables, the government is already having to intervene to buy back-up capacity, mainly from gas plant. How sure is Joe that you will be able to rely entirely on renewables in the future?

He recognises that there is a long way to go to realise his vision. He is very confident that 50% of electricity could come from renewables by 2050, but he recognises that, how far it is possible to go above that will depend on further technological progress as well as government investment and public and business responses to the need for flexibility.

Although the costs of renewables are coming down, the costs of managing intermittency are going up as the share of renewables increases.

If you are optimistic about what renewables can achieve and the potential for managing intermittency you have the option to spend P$4 billion in 2019 on subsidies for renewables. This will enable 75% of electricity to come from renewables in 2050. You may be somewhat more inclined to be optimistic about this if you are making a big push for the electrification of cars and homes, because these may contribute to the flexibility of the system. Otherwise renewables will remain restricted to no more than 50% of electric generation. However, if you have adopted the carbon tax, which already provides incentives for low-carbon energy, you can reduce the costs to P$3 billion.

Nuclear Power

Your next adviser is John. He has been listening to the last conversation. He gently points out that Rewop isn't necessarily the best place to rely on renewables, bearing in mind that peak electricity demand is in hours of darkness, that it is not unknown for still, windless weather to persist for several days, and the lack of any natural back-up such as hydro. He thinks you are going to need a more reliable source of energy. Nuclear power, he points out is the only fully commercialised source of reliable power available to a country like Rewop on a really large scale. John recognises that nuclear power has serious drawbacks but, he argues, they are not as bad as climate change. The people of Rewop will probably accept the need for nuclear power if the government puts a credible case.

You have some questions for John also. You have heard that construction of the latest nuclear power stations in developed economies has suffered serious cost overruns and delays. In contrast to renewables, nuclear power costs seem to be going up and not down.

John says that a Rewop project will benefit from the experience of these other reactors. In countries such as China, India, and South Korea, there are substantial nuclear programmes that are quite cost competitive. There are also a number of proposals for smaller, factory built, nuclear power stations that should not experience the same construction problems.

You have the option to spend P$2 billion in 2019 on a programme that will deliver an additional 50 units of electric power in 2050. If you are more optimistic about nuclear power, and believe that the costs could be brought down in an extended programme, you can spend P$4 billion to deliver 100 units. You can halve both of these costs to P$1 billion and P$2 billion if you have a carbon tax in place, because this will make nuclear power more competitive.

Over to You

You now need to work-up the presentation that you will make to the Prime Minister at your next meeting. There is, unfortunately, very little time. She expects you to tell her the following:

- What are the key judgements that you have made on how best to reduce Rewop's energy related CO_2 emissions?
- Is the target of reducing them to 120 million tonnes by 2050 realistic?
- If not, what level of emissions do you consider it realistic to aim for in 2050?
- How much will your plan cost energy consumers in 2019? She wants to confine this to P$10 billion or below if possible.
- How are the reductions in emissions to be achieved?
- What will be the components of Rewop's electricity supply in 2050. Can you assure her that it will be totally reliable?

Good luck!

Index

Printed in the United States
By Bookmasters